WOUND HEALING, FIBROSIS, AND THE MYOFIBROBLAST

WOUND HEALING, FIBROSIS, AND THE MYOFIBROBLAST
A Historical and Biological Perspective

GIULIO GABBIANI, MD, PhD
Department of Pathology and Immunology, Faculty of
Medicine, University of Geneva, Geneva, Switzerland

MATTEO COEN, MD, MD-PhD
Division of General Internal Medicine,
Department of Medicine, Geneva University Hospitals,
and Unit of Development and Research in Medical
Education, Faculty of Medicine, University of Geneva,
Geneva, Switzerland

FABIO ZAMPIERI, PhD
Department of Cardiac, Thoracic, Vascular Sciences and
Public Health, University of Padova Medical School,
Padova, Italy

ACADEMIC PRESS

An imprint of Elsevier

ELSEVIER

Library of Congress Cataloging-in-Publication Data
A catalog record for this book is available from the Library of Congress

British Library Cataloguing-in-Publication Data
A catalogue record for this book is available from the British Library

ISBN: 978-0-323-90546-6

For information on all Academic Press publications
visit our website at https://www.elsevier.com/books-and-journals

Publisher: Stacy Masucci
Acquisitions Editor: Katie Chan
Editorial Project Manager: Sam Young
Production Project Manager: Punithavathy Govindaradjane
Cover Designer: Matthew Limbert

Typeset by STRAIVE, India

Working together
to grow libraries in
developing countries

www.elsevier.com • www.bookaid.org

Cover image legend

Wound man, Pseudo-Galen,[a] Anatomia; WMS 290. The wound man of medieval anatomy is a figure illustrating various blows and lacerations to the human body by weapons such as clubs and knives. Credit: Wellcome Collection. Attribution 4.0 International (CC BY 4.0).

References

Hartnell, J. (2017). Wording the wound man. *British Art Studies.*

Petit, C. (2014). What does Pseudo-Galen tell us that Galen does not? Ancient medical schools in the Roman Empire. *Bulletin of the Institute of Classical Studies, Supplement*(114), 269–290.

Sudhoff, K. (1908). Der "Wundenmann" in Frühdruck und Handschrift und sein erklärender Text. Ein Beitrag zur Quellengeschichte des "Ketham". *Archiv für Geschichte der Medizin, 1*(5), 351–361.

[a] The many works assumed to be wrongly ascribed to Galen are referred as to pseudo-Galenic works (Petit, 2014). Among them the mid-15th century "Anatomia" (*liber uenerabilem Anotomeum [sic] secundum Galieleni [sic]*) from which the image of this "wound man" is taken. The Wound Man (or Wounded man) is a mysterious and interesting anatomical figure that can be found in various Medieval manuscripts. It illustrates a number of different injuries (e.g., blows, lacerations, sometimes also snake and insect bites) to which the human body can be prone (Hartnell, 2017). The German medicine historian Karl Sudhoff referred to the "Wound Man" as a "St. Sebastian distorted into a surgical grotesque" (*Chirurgisch-Groteske verzerrter St. Sebastian*) (Sudhoff, 1908).

Dedication

This book is dedicated, with love to:
Françoise
Adriano, Annamaria, Sara, Pietro, Emile
Norma e Girolamo

Contents

Preface I: The value of interdisciplinary collaboration in medicine and biology

The need for interdisciplinary cooperation to advance research and application of its results has long been asserted. This need is particularly felt in medicine and biology, due to the complexity of the objects studied and to the centrality of the health consequences that such research can produce. If the dialogue is often strongly sought after, more rarely this dialogue results in a concrete and valuable product.

This is the case of this book, which results from a multidisciplinary, collaborative approach between a historian of medicine, a clinician and a medical researcher, who together reconstruct the long path that led to the discovery of the nature and role of the essential components of the organization of the "living machine," first generally referred to as "fibers," then as various types of tissues, and finally individualized into specific structures, in particular the myofibroblast, which plays a central role in the processes that give the book its title, wound healing, and fibrosis.

One can enthusiastically welcome the collaboration between three authors representing the three fundamental components of medical thought and health care practice: history and philosophy, which reflect on the fundamental concepts of medical theory and practice, as well as their historical development; laboratory medicine, which experimentally studies and analyzes physiological and pathological structures and processes; and finally clinic, which brings this knowledge to the "bedside," allowing for treatment and health promotion.

We increasingly speak of "translational medicine," to indicate the necessary transition from the laboratory to the clinic and the population. But what is transmitted, in time, is also the history of a discipline or of a theory, and medicine itself is, after all, a historical science, for its ability to reconstruct the clinical history of each patient and to place in the right dimensions in time and space the development of diseases and epidemics. What is necessary to "translate," is therefore also the way in which knowledge and practice have developed over time. Even the fundamental concepts of medical knowledge, such as those of health and disease, have a necessary historical dimension. As suggested by the great historian of medicine Mirko D. Grmek, the concept of disease itself is historically determined, since "disease in general and all diseases in particular are concepts that do not immediately derive from our experience. They are explanatory models of reality and not of its constitutive elements. In short, diseases really exist only in the world of ideas. They interpret a complex empirical reality and therefore assume a certain medical philosophy, a system of pathological references." These references change over time, and the disease remains a notion determined by the historical context and cultural and ethical values, as for the tenuous boundary between normal and pathological, which varies over time depending on the possibilities of observation and identification of signs and symptoms, as by the social and cultural acceptance of what is "normal" or "pathological."

"Scientific medicine," as it was developed from the beginning of the 19th century with the contemporary origin of modern clinic and laboratory medicine, finds its point of origin in the anatomo-clinical work of Xavier Bichat (1771–1802) and in the historical and philosophical reflection of Philippe Pinel (1745–1821). In addition to identifying the places where medical knowledge, clinic and laboratory medicine are produced, scientific medicine establishes a new fundamental principle around which to build the explanatory structure of medicine. This principle is identified in the lesion, a "change in an organ" that must be sought and identified with all available means of analysis, in particular pathological anatomy, and fixed in a classification, in an "atlas." For Bichat, pathological anatomy is the medical science of the future, which will truly be able to provide medicine with a scientific basis, allowing the study of lesions of the body related to different diseases, without having to resort to "accessory sciences," such as chemistry or physics. It is in this way that is finally overcome the fracture, the dissociation between philosophy and practice, typical of classical medicine, which can be defined as the "schism of Galen." The scientific knowledge, anatomical, physiological, or chemical, was considered a philosophical knowledge, functional to the cultural formation of the doctor, but without a real influence on his practice, on his "know how." The search for universality, forms, and general causes of a phenomenon, was philosophical, while in medical practice it was only the experience of individuality that counted; the doctor could only propose explanations and local causes of an individual event.

The dissociation between knowledge and practice is overcome with the development of scientific medicine, which binds, thanks to a complex theoretical structure, the individual case to universal explanations. The medical practice, i.e., every single diagnostic or therapeutic act, must be based on scientific knowledge, and therefore on the general laws of "normal" and "pathological." The belonging of a definition or of a medical activity to a well-defined theoretical system, accepted on scientific and experimental basis, becomes the criterion of demarcation between scientific medicine and non-scientific empirical practice.

At the beginning of the 19th century, starting from clinical medicine and then more and more with laboratory medicine, which became equipped with increasingly powerful observation tools, structure and function become the object of analysis, through which abnormalities were identified and evaluated: the new definitions allowed to define the normal by opposition to the pathological. The comparison of a sick individual with others in a sufficiently large series, will allow perceiving the permanence of a specific lesion and the definition of the normality of a function or a structure.

The classification of diseases in nosological compartments and the definition of a specific lesion as an "abstract" cause of a specific disease set the goal for medicine to build a general science of disease, provided with the same epistemological structure of exact sciences. What distinguishes scientific medicine from pure empiricism or folk medicine is the capacity to know the specific and universal causes of disease.

Scientific medicine is based on the epistemological decision to base the explanation of disease and therapeutic intervention on the knowledge of the cause and effects of a specific injury, subordinating the individual case to general laws. From this moment on, medical practice, i.e., every single diagnostic or therapeutic act, must be based on scientific knowledge, and therefore strictly depends on the general laws of biology and pathology.

The topics dealt with in this book are particularly suitable to show how the notion of "specific lesion" plays a central role in scientific explanation and in medical practice and how the refinement of the description of anatomical structures and physiological functions allows to deepen and distinguish with increasing precision the organs involved and their pathological changes. These are "long-standing" themes, given that the treatment of wounds certainly has been the first application of medicine, from herbal pharmacology to surgery, as can be seen from many scenes appearing on Greek and other ancient vases. Homer celebrates the exploits of the two brothers Macaon and Podalirius, physicians and surgeons capable of preventing epidemics and of treating ulcers and wounds. Moreover, the history of the treatment of acute and chronic wounds is present in ancient Egypt, as documented by many medical papyri, with collections of herbals, surgical and magical remedies.

After the origin of modern anatomy in the 16th century, the concept of fiber becomes the explanatory element of health and disease, differentiating into more and more specific tissues, until today with the progress, from the second half of the 19th century and especially in the second half of the 20th century, of microscopy, imaging techniques and study of cells. The theme, on the other hand, touches two of the most important objects of biological and medical research, i.e., tissue regeneration on the one hand and pathological proliferation on the other.

It is the specificity of the structure and of the lesion, which enables scientific explanation, as well as cure. A scientific explanation in medicine can indeed be considered complete when it is possible to establish a close connection, circular in nature, between disease specificity, lesion specificity, and cause specificity. A decisive role in the construction of the epistemological structure of scientific medicine has been played by the possibility of recognizing a specific pathological lesion, identifying its location, and linking it to a specific disease. Although the identification of a lesion has made use from the origin of classical medicine of all senses, such as touch, taste, useful to diagnose diabetes from the sweet taste of urine, smell capable of perceiving miasmas and putrefaction, hearing, as in auscultation, especially after the introduction of the stethoscope by Laennec, it is mainly the sight that allows to observe the lesions and especially to prepare a graphic representation, for subsequent research, diagnosis, and teaching.

It is indeed the concept of specificity of an organ and a lesion that allows establishing a strong theoretical link between the disciplinary domains discussed extensively in this book: nosology, which aims at a classification of disease, clinic, which aims at diagnosing and treating a specific disease, and pathological anatomy, which isolates and illustrates a lesion.

In medicine, the introduction of the experimental method, whether carried out in a laboratory, in a clinic or on the field, marks the transition from a vague and indefinite causality, in which multiple and unpredictable causes are at stake, to the establishment of a specific cause responsible for a given physiological or pathological phenomenon. The apparently complex and variable plans of biological systems are brought back to a determinism that can be complex, but always specific and regular. Even the objectives of treatment become much more precise, thus linearly increasing the possibility of error, given that the "target" is more limited; similarly, ethical, professional, legal, and criminal responsibilities become more precise helping the correct identification of the causes of disease and death.

In this way, medicine acquires a dual nature: on the one hand, it remains a science that requires a broad structure of laws, concepts, and theories that "frame" and allow the case to be "treated" theoretically. On the other hand, the attention is directed toward the specific patient and the phenomena perceived practically (clinical signs, laboratory data, pathological findings, individual history, etc.) become nodes of a theoretical network that builds an image of reality. The comparison between the "clinical picture" and the theoretical framework will allow to recognize a given specific disease and look for its equally specific and necessary cause.

The way in which this book analyzes the progressive clarification of the concepts of fibrosis and myofibroblast is a valuable and fruitful example of how the identification of specific organs and cells, their functions and alterations, equally specific, allows to understand the nature of physiological and pathological processes, thus making a fundamental contribution to the development of medical knowledge and of the capacity to treat diseases. We can only be grateful to the three authors, who could overcome the narrowness of disciplinary boundaries to propose an integrated interpretation of a beautiful chapter in the history of biology and medicine.

Bernardino Fantini
Honorary Professor of History of Medicine,
University of Geneva, Geneva, Switzerland

Preface II: Myofibroblasts: 50 years and aging well

The year 2021 marks the 50th anniversary of the discovery of the myofibroblast. Regardless of their advanced age, myofibroblasts have never been more popular to study, for a variety of reasons: (1) Myofibroblasts have an undeniable clinical impact by promoting tissue contractures in physiological healing and pathologies associated with organ fibrosis (Henderson, Rieder, & Wynn, 2020). Fibrosis becomes more prevalent in our aging population where individuals live longer than any previous generation and have the time and lifestyle to develop fibrosis risk conditions, such as diabetes and hypertension. (2) Myofibroblasts and their fibroblast relatives are positioned at the crossroads of clinical and basic research. Both cells are frequently interrogated culture models to answer fundamental questions in cell mechanobiology and molecular cell biology. (3) New technologies, such as cell lineage tracing, single cell RNA sequencing (scRNAseq) and multiplexing tools now allow systematic analysis of fibroblast and myofibroblast functional and phenotypic heterogeneity. This timely book will touch upon many of these aspects of myofibroblast biology in health and disease.

Despite, or possibly because of five decades of research, the term "myofibroblast" can have different meanings to different researchers and—facing such a debate—it is advisable to return to the roots of a name. Myofibroblasts were originally discovered in healing rat wound granulation tissues, and the name was chosen to reflect their sharing of morphological features of contractile smooth muscle cells (myo) and features attributed to extracellular matrix (ECM)-producing fibroblasts. The first description of myofibroblasts at the ultra-resolution level was published in Experientia by Gabbiani and coworkers, stating that tissue myofibroblasts "showed the cytologic structure regarded as 'typical' […] (i.e. numerous cisternae of rough endoplasmic reticulum and many mitochondria)" and later exhibited "bundles of packed fibrils resembling those of smooth muscle" (Gabbiani, Ryan, & Majno, 1971). This morphological assessment was followed by the first functional study demonstrating contraction of myofibroblast-populated tissues upon stimulation with smooth muscle agonists. The latter study was also the first introducing the new name: "However, our results support the conclusion that fibroblasts, under certain conditions, are capable of modulating toward a cell type that is structurally and functionally close to smooth muscle; for these cells the name 'myofibroblast' may be appropriate." (Majno, Gabbiani, Hirschel, Ryan, & Statkov, 1971). It is passed on that "fibroplast" was also discussed among the authors of these two seminal studies, but that label did not stick.

Ambiguity in using the term myofibroblast as a cell discriminator can arise because both fibroblasts and smooth muscle cells are phenotypically plastic cells that can exist in different states and are difficult to describe by unique identifiers (Arnoldi, Chaponnier, Gabbiani, & Hinz, 2012; Mascharak, desJardins-Park, & Longaker, 2020; Shaw & Rognoni, 2020). Fibroblasts, in the most general sense, are connective tissue cells with spindle-shape morphology and the ability to produce connective tissue substance, i.e., collagen (Virchow, 1858; Ziegler, 1895). However, this loosely defined morphology, function, and location not only fits bona fide fibroblasts but also pericytes, perivascular cells, mesenchymal stromal/stem cells (MSC), and fibro-adipogenic progenitors (FAP) (Lemos & Duffield, 2018). While being superficial lookalikes, these distinct populations perform specific functions in maintaining tissue homeostasis and regulating local biological processes in the healthy adult organism. Yet, they all have the capacity to become activated and converge to the myofibroblast phenotype in conditions of tissue injury and/or chronic inflammation (Pakshir et al., 2020). The potential of different precursors activating onto myofibroblasts is directly related to the principle function of this phenotype to acutely repair tissue injuries by rapidly establishing mechanically resistant collagen scar—wherever and whenever the injury occurs (Bochaton-Piallat, Gabbiani, & Hinz, 2016; Hinz, McCulloch, & Coelho, 2019).

As the hallmark of myofibroblast activation, neoformation of actin-myosin bundles in vivo or of the equivalent stress fibers in vitro principally satisfies the demand on myofibroblast morphological criteria but is not useful for routine immuno-histochemical identification and/or diagnosis. Agents that discriminate between non-polymerized and filamentous stress fiber actin, such as fluorescently labeled phalloidin, are not compatible with organic solvents used in standard histology (Smith-Clerc & Hinz, 2010). Moreover, all cultured fibroblastic cells form stress fibers upon contact with conventionally stiff culture surfaces and are thus technically myofibroblasts. To sensitize fibroblast researchers to the fact that they are studying myofibroblasts in culture, Tomasek and coworkers introduced the term proto-myofibroblast (Tomasek, Gabbiani, Hinz, Chaponnier, & Brown, 2002). The proto-myofibroblast is considered a stress fiber forming pre-stage of the 'differentiated' myofibroblast, which is discriminated by the neoexpression of the actin isoform α-smooth muscle actin (α-SMA) in stress fibers. Incorporation of α-SMA into stress fibers enhances the contractile activity of proto-myofibroblasts and removal of α-SMA from this location using specific competitive peptides represents a potential strategy to counter tissue contractures (Chaponnier et al., 1995; Hinz, Gabbiani, & Chaponnier, 2002).

The development of actin isoform-specific antibodies was another milestone in myofibroblast biology leading to the discovery that myofibroblasts, unlike most of their precursors, neoexpress α-SMA in healing wounds (Darby, Skalli, & Gabbiani, 1990; Skalli et al., 1986). Expression of α-SMA has since become the most commonly and conveniently used molecular marker of myofibroblasts in cell culture and in fibrotic tissues (Chaponnier & Gabbiani, 2016). Other frequently used myofibroblast identifiers are the extradomain-A (ED-A) splice variant of fibronectin (Serini et al., 1998) and periostin (Nikoloudaki, Creber, & Hamilton, 2020) in the ECM, and platelet-derived growth factor receptor beta (PDGFRβ) (Henderson et al., 2013), α11β1 integrin (Schulz et al., 2018) and cadherin-11 on the plasma membrane (Lodyga et al., 2019). Although these markers are enriched at various levels depending on the context, none of them are unique to myofibroblasts, again owing to the heterogeneous progeny of these cells and different environmental context.

Another open question that will be discussed in this book is that of myofibroblast fate. Transient activation of fibroblastic cells is crucial and beneficial to repair acute tissue damage whereas their persistent activation characterizes and drives fibrosis. The factors deciding whether myofibroblasts persist or resolve are still unclear. In addition to their identity (i.e., inherent fate), local environment critically determines myofibroblast function and persistence, including mechanical factors like tissue stiffness and strain, chemical factors like TGF-β1, and cellular environment like inflammatory macrophages (Desmouliere, Geinoz, Gabbiani, & Gabbiani, 1993; Pakshir & Hinz, 2018). As part of a fibrotic positive feedback loop, myofibroblasts generate the very environment that promotes their formation: myofibroblast contractile activity stiffens the ECM, they produce and activate pro-fibrotic TGF-β1 from the ECM, and instruct other cells to perform pro-fibrotic actions (Lodyga & Hinz, 2020). Consequently, interfering with these factors and the according myofibroblast reception and transduction mechanisms offers targets for therapeutic strategies that can stop, control, and possibly even revert dysregulated healing. The initial conception of a terminally differentiated myofibroblast may not apply to all myofibroblast states and in vivo evidence is accumulating that myofibroblast de-activation is possible (Jun & Lau, 2018). Nevertheless, apoptosis and senescence are frequently observed terminal states of the α-SMA-positive myofibroblast and whether myofibroblast prevail in fibrosis or disappear after acute repair may in fact be regulated by acquired resistance to death (Desmoulière, Redard, Darby, & Gabbiani, 1995; Hinz & Lagares, 2020).

Only with a thorough mechanistic understanding of myofibroblast function and identity, we will be able to devise-specific clinical strategies that target myofibroblasts. The inclined reader of this book will find that the stage is now well prepared for novel therapeutic concepts to control myofibroblast activation, activity, and survival in health and disease.

Boris Hinz

Laboratory of Tissue Repair and Regeneration, Faculty of Dentistry, University of Toronto, Toronto, ON, Canada

References

Arnoldi, R., Chaponnier, C., Gabbiani, G., & Hinz, B. (2012). Heterogeneity of smooth muscle. In J. Hill (Ed.), *Muscle: Fundamental biology and mechanisms of disease* (pp. 1183–1195). Elsevier Inc.

Bochaton-Piallat, M. L., Gabbiani, G., & Hinz, B. (2016). The myofibroblast in wound healing and fibrosis: Answered and unanswered questions. *F1000Res, 5*.

Chaponnier, C., & Gabbiani, G. (2016). Monoclonal antibodies against muscle actin isoforms: Epitope identification and analysis of isoform expression by immunoblot and immunostaining in normal and regenerating skeletal muscle. *F1000Res, 5*, 416.

Chaponnier, C., Goethals, M., Janmey, P. A., Gabbiani, F., Gabbiani, G., & Vandekerckhove, J. (1995). The specific NH2-terminal sequence Ac-EEED of alpha-smooth muscle actin plays a role in polymerization in vitro and in vivo. *The Journal of Cell Biology, 130*, 887–895.

Darby, I., Skalli, O., & Gabbiani, G. (1990). Alpha-smooth muscle actin is transiently expressed by myofibroblasts during experimental wound healing. *Laboratory Investigation, 63*, 21–29.

Desmouliere, A., Geinoz, A., Gabbiani, F., & Gabbiani, G. (1993). Transforming growth factor-beta 1 induces alpha-smooth muscle actin expression in granulation tissue myofibroblasts and in quiescent and growing cultured fibroblasts. *The Journal of Cell Biology, 122*, 103–111.

Desmoulière, A., Redard, M., Darby, I., & Gabbiani, G. (1995). Apoptosis mediates the decrease in cellularity during the transition between granulation tissue and scar. *American Journal of Pathology, 146*, 56–66.

Gabbiani, G., Ryan, G. B., & Majno, G. (1971). Presence of modified fibroblasts in granulation tissue and their possible role in wound contraction. *Experientia, 27*, 549–550.

Henderson, N. C., Arnold, T. D., Katamura, Y., Giacomini, M. M., Rodriguez, J. D., McCarty, J. H., et al. (2013). Targeting of alphav integrin identifies a core molecular pathway that regulates fibrosis in several organs. *Nature Medicine, 19*, 1617–1624.

Henderson, N. C., Rieder, F., & Wynn, T. A. (2020). Fibrosis: From mechanisms to medicines. *Nature, 587,* 555–566.

Hinz, B., Gabbiani, G., & Chaponnier, C. (2002). The NH2-terminal peptide of alpha-smooth muscle actin inhibits force generation by the myofibroblast in vitro and in vivo. *The Journal of Cell Biology, 157,* 657–663.

Hinz, B., & Lagares, D. (2020). Evasion of apoptosis by myofibroblasts: A hallmark of fibrotic diseases. *Nature Reviews Rheumatology, 16,* 11–31.

Hinz, B., McCulloch, C. A., & Coelho, N. M. (2019). Mechanical regulation of myofibroblast phenoconversion and collagen contraction. *Experimental Cell Research, 379,* 119–128.

Jun, J. I., & Lau, L. F. (2018). Resolution of organ fibrosis. *The Journal of Clinical Investigation, 128,* 97–107.

Lemos, D. R., & Duffield, J. S. (2018). Tissue-resident mesenchymal stromal cells: Implications for tissue-specific antifibrotic therapies. *Science Translational Medicine, 10,* eaan5174.

Lodyga, M., Cambridge, E., Karvonen, H. M., Pakshir, P., Wu, B., Boo, S., et al. (2019). Cadherin-11-mediated adhesion of macrophages to myofibroblasts establishes a profibrotic niche of active TGF-beta. *Science Signaling, 12.*

Lodyga, M., & Hinz, B. (2020). TGF-beta1—A truly transforming growth factor in fibrosis and immunity. *Seminars in Cell & Developmental Biology, 101,* 123–139.

Majno, G., Gabbiani, G., Hirschel, B. J., Ryan, G. B., & Statkov, P. R. (1971). Contraction of granulation tissue in vitro: Similarity to smooth muscle. *Science, 173,* 548–550.

Mascharak, S., desJardins-Park, H. E., & Longaker, M. T. (2020). Fibroblast heterogeneity in wound healing: Hurdles to clinical translation. *Trends in Molecular Medicine, 26,* 1101–1106.

Nikoloudaki, G., Creber, K., & Hamilton, D. W. (2020). Wound healing and fibrosis: A contrasting role for periostin in skin and the oral mucosa. *American Journal of Physiology. Cell Physiology, 318,* C1065–C1077.

Pakshir, P., & Hinz, B. (2018). The big five in fibrosis: Macrophages, myofibroblasts, matrix, mechanics, and miscommunication. *Matrix Biology.*

Pakshir, P., Noskovicova, N., Lodyga, M., Son, D. O., Schuster, R., Goodwin, A., et al. (2020). The myofibroblast at a glance. *Journal of Cell Science, 133.*

Schulz, J. N., Plomann, M., Sengle, G., Gullberg, D., Krieg, T., & Eckes, B. (2018). New developments on skin fibrosis—Essential signals emanating from the extracellular matrix for the control of myofibroblasts. *Matrix Biology, 68-69,* 522–532.

Serini, G., Bochaton-Piallat, M. L., Ropraz, P., Geinoz, A., Borsi, L., Zardi, L., et al. (1998). The fibronectin domain ED-A is crucial for myofibroblastic phenotype induction by transforming growth factor-beta1. *The Journal of Cell Biology, 142,* 873–881.

Shaw, T. J., & Rognoni, E. (2020). Dissecting fibroblast heterogeneity in health and fibrotic disease. *Current Rheumatology Reports, 22,* 33.

Skalli, O., Ropraz, P., Trzeciak, A., Benzonana, G., Gillessen, D., & Gabbiani, G. (1986). A monoclonal antibody against alpha-smooth muscle actin: A new probe for smooth muscle cell differentiation. *The Journal of Cell Biology, 103,* 2787–2796.

Smith-Clerc, J., & Hinz, B. (2010). Immunofluorescence detection of the cytoskeleton and extracellular matrix in tissue and cultured cells. *Methods in Molecular Biology, 611,* 43–57.

Tomasek, J. J., Gabbiani, G., Hinz, B., Chaponnier, C., & Brown, R. A. (2002). Myofibroblasts and mechano-regulation of connective tissue remodelling. *Nature Reviews. Molecular Cell Biology, 3,* 349–363.

Virchow, R. (1858). *Die Cellularpathologie in ihrer Begründung auf physiologische und pathologische Gewebelehre.* A. Hirschwald.

Ziegler, E. (1895). *General pathology; or, the science of the causes, nature and course of the pathological disturbances which occur in the living subject.* New York: W. Wood and Company.

Preface III

This book represents a rare case of a strict collaboration between a historian of medicine (FB), a physician involved in clinical practice (MC), and a physician and researcher who has been the protagonist of the discovery of the nature and role of the myofibroblast (GG), which is the main focus of this historical and biological perspective. At the same time, it represents a likewise rare case of combination between ancient and contemporary history.

The aim of this book is to present the reader with the historical background in which the main concepts that are the object of our work, i.e., the fibrosis and the myofibroblast, have emerged. Moreover, it wishes to illustrate the biological significance of these concepts for diagnosis, clinical management, and possible therapeutic strategies of several widespread diseases, such as fibromatoses and organ fibroses. We hope that these aims will meet, at least in part, the interest of readers with different scientific and clinical backgrounds.

Giulio Gabbiani
Matteo Coen
Fabio Zampieri

Acknowledgments

We are particularly grateful to Professors Bernardino Fantini, University of Geneva, Geneva, Switzerland, and Boris Hinz, University of Toronto, Ontario, Canada for writing a preface to our work.

We thank the numerous students, post-docs and visitors, who, at the Department of Pathology, University of Geneva, have contributed throughout the years to the evolution and definition of the myofibroblast concept, in particular: Christine Chaponnier, Omar Skalli, Olivier Kocher, Alexis Desmoulière, Marie-Luce Bochaton-Piallat, Guido Serini, Sophie Clement, Elisabeth Rüngger-Brändle, Boris Hinz, Robert B. Low, Ian Darby, Anna Fagotti, Rita Pascolini, Paolo Leoncini, H. Paul Ehrlich, Ottavio Cremona, Annette Schmitt-Gräff, Istvan Hüttner, Yusuf Kapanci, Walter Schürch.

We thank the following Journals: Advanced Drug Delivery Reviews, Drug Discovery Today, EMC-Podología, Experientia, Journal of Cell Biology, Laboratory Investigation, The American Journal of Pathology, The Lancet, The International Journal of Cancer, publishers: Birkhäuser Verglag, Elsevier Masson SAS, Rockefeller, University Press, Springer Nature, and Institutions: Belle Arti e Paesaggio di Roma, Bibliothèque nationale de France, Governorate of Vatican City State, Directions of Museums, Libraries & Collections at King's College London, Ministero per i beni e le attività culturali e per il turismo, Soprintendenza speciale Archeologia, Belle Arti e Paesaggio di Roma, Museu Nacional d'Art de Catalunya, Barcelona, RMN-Grand Palais (musée du Louvre), The Penn Museum, The Wellcome Collection, for allowing reproduction of the figures. We thank the "Fond G. Gabbiani," Marie-Luce Bochaton Piallat and Anita Hiltbrunner, University of Geneva, for the help in obtaining figure reproductions.

We are particularly thankful to the following colleagues, who shared their pictures with us: Dr. Rosa Dinares Solà, (Hospital General de Catalunya, Barcelona, Spain), Prof. Milton Nuñez, (University of Oulu, Finland), Dr. Michael Papaloizos, (Centre de Chirurgie et de Thérapie de la Main, Geneva, Switzerland), Dr. Dan Adler, (Division of Pulmonary Diseases, Geneva University Hospitals, Geneva, Switzerland) Prof. Laura Rubbia, (Division of Clinical Pathology, Geneva University Hospitals, Geneva, Switzerland); and to Prof. Giovanni Gallo for his picture of Moses' hand.

Finally, we deeply thank Dr. Telli Faez for her precious advices and support.

CHAPTER 1

Fibrous tissue, fibrosis, and the fibroblast

Fibrosis is defined as an abnormal formation and accumulation of fibrous tissue or, according to a terminology common in the medical language of the 19th century, is a *"fiber hyperplasia."* The role of the fibroblast in this phenomenon was appreciated only during the 20th century as we shall see below.

The concept of fiber, from which fibrosis and fibroblast are derived, is very old and has a complex history. Joseph Hyrtl (1810–94), in his history of the anatomical language, stated that *"no other anatomical definition had so remarkably changed his meaning as fiber"* (Hyrtl, 1884). Fiber derives from the Latin word *fibra*, indicating in agriculture a filament, a rut or a tree (Marcovecchio, 1993). The Latin term derives from vis and possibly from the Greek *is*, meaning *"strength"* and referring to the plant portion that resists to external injuries. In the classic and medieval medical language, the term *fibra* was equivalent to the present term "lobe" (from the Latin *lobus*) as seen in Roman medical books such as the *De medicina* by Aulus Cornelius Celsus (c. 25 BC-c. AD 50); discussed in Skinner (1970, p. 174) and Marcovecchio (1993, pp. 360–361).[a] Hyrtl (1884, p. 147) suggested that *fibra* was used with the meaning of "extremity," indicating the distal parts of the liver or the lungs. At the beginning of the 16th century the term *lobus* started to be used currently instead of *fibra* for the liver and lungs (Marcovecchio, 1993, p. 511). Alessandro Benedetti (1450–1512), anatomist and humanist at Padua University, was the first to use *fibra* with both meanings: in his book *"Anatomice sive historia corporis humani"* ("Anatomy, or the history of the human body") (Benedetti, 1502), in addition to the traditional use in the description of liver and lungs, he applied the term in the

---

[a] Celsus was a Roman encyclopedist who wrote an encyclopedic compendium of the whole scientific and technical knowledge of his time, in which the section on medicine was probably one of the most important. His *De medicina* was virtually unknown in the West until the beginning of the 15th century, when it was rediscovered and then became the principal source of Renaissance medical terminology (Zampieri, 2016, p. 102).

Wound Healing, Fibrosis, and the Myofibroblast
https://doi.org/10.1016/B978-0-323-90546-6.00006-X

description of blood clot, the thin tubes in which ends the trachea and small vessels (Ferrari, 1996). However he employed the term *stamen* for the tissue we now call connective tissue (Ferrari, 1996).

The modern use of *fibra*, as an elongated and thin structure, applied to muscle, nerves, ligaments, tendons, and small vessels, begins with Andreas Vesalius (1514–64) and Gabriele Falloppia (1523–62). Such meaning will be further elaborated on the basis of the anatomical tissue characterization, based not only on the macroscopic, but also on microscopic investigation, during the 18th and the 19th centuries. In *De humani corporis fabrica* Vesalius speaks of fibers as constituents of cerebral "*membranes*" (Vesalius, 1543). He used the same term in the description of muscles, suggesting that muscles are formed through the assembly (*congressus*) of *fibrae* of nerves and ligaments, from which muscle fibers depart (Vesalius, 1543). Moreover, Vesalius describes three different *fibrae* that contribute to the vein structure (Vesalius, 1543). In the book *Epitome*, a short version of *Fabrica* published in 1543, we find the description of elementary components of the human body, i.e., bones, cartilages, ligaments, fibers, membranes, muscles and fat tissue (Vesalius, 1543).

Falloppia can be considered a precursor of the notion of tissue. In some lectures published by his student Volcher Coiter (1534–1600) in 1575 he discussed what he called "*basic structures*" (*partes similares*) that make up the organs and correspond to the modern concept of "*tissue*" (Falloppia, 1575). Here we cannot describe in detail this concept; it suffices to say that, according to Falloppia, such basic structures were composed of *fibrae*. These were exerting three main functions: (1) structural, representing the main constituent of the solid portions of the body; (2) dynamic, constituting the *fibra carnea* and the *fibra cartilaginea* they were facilitating the voluntary and spontaneous movements; (3) communicative, allowing through their organization in the space the formation of tubes or pores through which "*nourishing fluids*" or "*spirits*" of the body could circulate, according to the physiological concept inspired by Galen's (AD 129–216) theories still accepted at that time (Ongaro, 1981). This view had probably been anticipated by the French physician Jean Fernel [1497–1558], a visionary precursor personality, to whom we owe the term "*physiology*," who proposed that organs and tissues were composed of a network of fibers (Berg, 1942). A very similar theory can be found at the end of the 17th century in the popular anatomy textbook by the Scottish physician James Keill (1673–1719) "*The Anatomy of the Human Body, abridged*."

In the introduction, Keill suggests that the human body is composed of different types of *"fibers,"* some soft and elastic, some cavernous and some filled by small cells; this could be verified by microscopic analysis. Microscopic anatomy had started to become a tool for medical investigation during the 17th century, in particular thanks to the pioneering work of Marcello Malpighi (1628–94) (Zampieri, 2016, pp. 67–140). Keill writes:

> [...] *all the parts are made up of threads, or fibers, of which there be different kinds; for there are some soft, flexible, and a little elastick; and there are either hollow, like small pipes, or spongeous, and full of little cells, as the nervous and fleshy fibres; others there are more solid and flexible, but with a wrong elasticity or spring, as the membranous and cartilaginous fibres; and a third sort are hard and inflexible, as the fibres of the bones. Now of all these, some are very sensible, and others are destitute of all sense; some so very small as not to be easily perceived; and others, on the contrary, so big as to be plainly seen. And most of them, when examined with a microscope, appear to be composed of still smaller membranes.*
>
> **(Keill, 1698, p. 2)**

According to the mechanistic approach used in the 17th and 18th centuries, the concept of fiber was instrumental for the understanding of the function of organs and tissues, which were considered to be composed of fibers exhibiting different physical properties, such as elasticity and flexibility; these differences could explain in physical and mathematic terms the various activities of organs. Albrecht von Haller (1708–87) in his book *Elementa physiologiae* introduced the view that the fiber represents for physiology what represents the line for geometry, i.e., the fundamental structural element ("*Fibra enim phsiologo id est, quod linea geometrae*") (von Haller, 1757, p. 2). The fiber became the basic element of life and this role was conceptually transferred gradually to the cell during the 19th century as we shall see below (Berg, 1942). During the 18th century several terms used until then were created from *fibra,* such as fibrin, suggested by Haller to define the basic element of blood clot and fibril from the French "*fibrille,*" indicating a small constituent of fibers.

At the end of the 18th century and the first half of 19th century anatomical research was concentrated on the characterization of "membranes," i.e., the tissues composing the various organs. In this period, we see the emergence of the concept of connective tissue, which obviously is of paramount interest for our work. In this respect John Hunter (1728–93) deserves a particular mention: allegedly before Bichat, about whom we shall speak below, he had the intuition that organs were composed of different tissues and that similar tissues could react pathologically in the same way even when located in different organs. Hunter suggested that inflammation elicited in tissues three different reactions: adhesions, pus formation (suppuration) and ulceration. Even more significant was the assumption that each of these reactions was typical of a specific tissue: adhesions were typical of serous and cellular tissues, pus formation of mucous and adipose tissues and ulcers of cutaneous and connective tissues (Maulitz, 1987, pp. 114–117).

According to Hunter (1794): "*The cellular membrane, free from the adipose, appears to be more susceptible of the adhesive inflammation than the adipose membrane, and much more readily passes into the suppurative. [...] Ulceration [...] does not so readily take place in those parts as it does in the common connecting membrane*" (p. 206).

Marie François Xavier Bichat (1771–1802), the famous French anatomist considered the father of the disciplines now named as histology and histopathology, was the first to characterize "*fibrous*" tissue; his work has furnished the basis for the understanding the features of what now is called connective tissue. In his well-known "*Traité des Membranes*" Bichat described three types of "*simple*" membranes: mucosae, serosae, and fibrous membranes. Fibrous membranes were "dry," composed of white fibers similar to the basic component of tendons; they made up periosteum, dura mater, sclera, cavernous body capsule, joint capsules, and tendon sheaths. In his monumental "*Anatomie Générale*" whose first volume was published in 1802 and the next two volumes appeared in 1812, after his death, Bichat presented a much more complex description identifying 21 fundamental tissues of the human body including fibrous tissue that composes periosteum, capsules, fibrous sheaths, aponeuroses, tendons and ligaments (Bichat, 1812, pp. 145–210). Fibrous tissue has a structure with intermingled layers; it is hard, not elastic, neither sensitive nor contractile, and exerts essentially a static function connecting other tissues. It surrounds organs that are not subject to alternating dilatations and contractions, such as the spleen or testes, in contrast to stomach or lungs.

Other studies contributed to the histological and pathological characterization of "fibrous tissue" during the first half of the 19th century: the German physiologist Johan Müller (1801–58) started to use the term "*connective tissue*" around 1830 (Paladino, 1887; Robb-Smith, 1954; p. 131 and 16, respectively). It should be noted that during this period the histological approach proposed by Bichat was applied only rarely, while most pathologists remained limited to the macroscopic approach. For example, in the first volume of the "*Manual of General, Descriptive and Pathological Anatomy*," the German pathologist Friedrich Meckel (1781–1833) stated that histology was less useful than the traditional macroscopic perspective in the study of normal and pathological anatomy (Meckel, 1838, p. 27). However, Meckel described some observations concerning structures with fibrous features. He writes (Meckel, 1838, p. 22):

> The organic form [...] presents two points of view: 1st, its intimate composition, the texture of its parts; 2nd, the external composition, the structure or the form. [...] As regards texture, the component parts may be reduced to others more simple, which in their turn differ from each other, in their degree of simplicity, and may, therefore, be divided into proximate and remote. The remote constituent parts of the organic form are finally reduced to two, of which one appears constantly under a given form which is not the case with the other, although this is equally susceptible of figure. These parts are the globules and a coagulated or coagulable substance.

The "*globules*" were allegedly cells, since he described globular bodies with various forms and dimensions according to the organ, including blood; it should be noted that at that time it was not established that cells were the fundamental constituents of all tissues. Meckel proposed that these two components combined make up "fibrous" or "lamellar" structures according to their proportions and their special organization:

> These two remote constituent parts, the globules and the coagulable liquid, produce, either the second alone, or both combined, two principal forms: in the first, the length much exceeds the other dimensions; in the second, it is more nearly equal to the breadth, although they both exceed the thickness. The first form is called fibrous, the second, the laminar: the fibrous form belongs usually only to the coagulated liquid, which is sometimes changed into fibers, even without the globules, as in the bones, tendons, &c.

(Meckel, 1838, pp. 24–25)

These components, once assembled, are forming tissues and organs and eventually are participating to the formation of pathological products. It should be noted that one of the few textbooks of pathology that followed entirely the ideas of Bichat was "*Lectures on the Morbid Anatomy of Serous and Mucous Membranes*" by the British Thomas Hodgkin (1798–1866), which unfortunately did not consider fibrous membranes.

Bichat had reported that fibrous tissue responds stereotypically to solicitations irrespective of the stimulus, even in pathological situations. This assumption was practically forgotten during the 19th century and reappeared in the 20th century.

The cellular theory, proposed by Theodore Schwann (1810–82) a student of Müller, represented a paradigmatic change about the nature of the elementary tissue components. According to Schwann even fibrous tissue was composed of cells that by coming together were forming fibers. In the book "*Microskopische Untersuchungen über Uebereinstimmung in der Structur und der Wachstum der Tiere und Pflantzen*" ("Microscopic studies on the similarity in the structure and growth of animals and plants") Schwann presented a microscopical drawing (Fig. 1.1) of connective tissue in which the cells were called "*fasernzellen*" (i.e., fibrous cells) (Schwann, 1839, p. 269). As it is well known, the cellular theory was popularized in the medical field by Rudolph Virchow (1821–1902) allowing the birth of what is called "cellular pathology." However, few years before the appearance of the famous

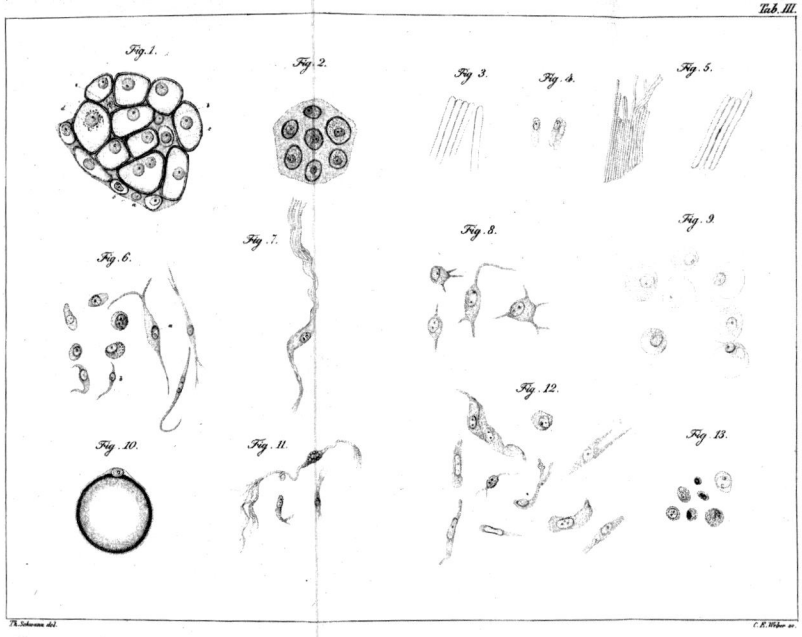

Fig. 1.1 Illustration of the different kinds of tissues and cells individuated by Theodore Swann, from the Table II of his *Mikroskopische Untersuchungen* ("Microscopic examinations") (Schwann, 1839). "Fibers" are represented in n. 5 with the following caption: "*Fasern, welche die substantia propria des menschlichen Zahns zusammensetzen, durch zweitägige Maceration mit verdünnter Salzsäure isolirt*" ("Fibers composing the substantia propria of the human tooth isolated by maceration with dilute hydrochloric acid for two days."). The n. 6 is defined by him as "faserzellen," that is, fibrous cell. This is the original caption: "*Faserzellen aus dem unter den Hautmuskeln des Halses liegenden Zellgewebe eines 7" langen Schweine-fötus*" ("Fiber cells from the cellular tissue underlying the skin muscles of the neck of a 7" long porcine fetus"). *(Reproduced with permission from Schwann, T. (1839). Mikroskopische Untersuchungen über die Uebereinstimmung in der Strucktur und dem Wachsthum der Thiere und Pflanzen. Berlin: Sander; Wellcome Collection. Attribution 4.0 International (CC BY 4.0).)*

"*Cellularpathlogie*" (Virchow, 1858), two remarkable pathological atlas were published, respectively, by Julius Vogel [1814–80], another student of Müller, "*Icones histologiae patologicae*" (Vogel, 1843) and by the German pathologist Herman Lebert (born Hermann Levay ou Lewy; 1813–78), "*Physiologie Pathologique*" (Lebert, 1845). See also Hajdu (2004) Lebert's atlas has been defined "*one of the most magnificent in the rich tradition of diagnostic pathology*", but "*...almost fallen into oblivion, being unknown to most contemporary workers*" (Pickel, Reich, Winter, & Young, 2009). Born in Breslau, then in Prussia (now Wrocław in Poland), he spent his academic life between France (Paris), Switzerland (Zurich), and Prussia (Breslau); he also worked together with the founder of experimental physiology in Geneva Jean-Louis Prévost (1838–1927). As a physician he mainly worked in the village of Bex (where he died), in the Swiss thermal baths of Lavey-Morcles (in the Swiss Canton of Vaud), and in Paris (Pickel et al., 2009). He served also as an army doctor in the Swiss civil war of the Sonderbund (in German, *Sonderbundskrieg*), in 1844, as well as in the Austro-Prussian war in Breslau in 1866 (Kaiser, 2010). A real erudite, his interests spun well beyond pathology, with publications on the benefits of thermal waters, as well as on botany and entomology (in particular on the water mites, Hydrachnidia) (Kaiser, 2010; Pickel et al., 2009).

In this last work are reported observations concerning diseases characterized by fibrous tissue deposits. In addition, Table III of this book shows what the author names "globules fibro-plastiques" (that is, "fibro-plastic globules") (Lebert, 1845; p. 23); see Figs. 1.2 and 1.3. Moreover, in the next table Lebert illustrates what he calls "intermediate forms between fibro-plastic globules and fibers" that are named "fusiform globules or bodies," with or without nucleus, and defined as "molecular components of the intercellular substance" (Lebert, 1845, p. 24). However, Lebert does not specify the pathological tissue analyzed. Finally, the term "fibro-plastic globules" is utilized by Lebert to characterize the cellular components of several tumors defined "fibro-cellular" (Lebert, 1845, p. 24); obviously, this complicates the interpretation of these terms.

With the book "Cellular Pathologie" by Virchow begins an interesting and fruitful debate on the nature of connective tissue and in particular on the origin of extracellular matrix. Virchow proposed a classification of tissues according to new criteria: (1) "cellular" tissue, composed of cells in contact with one other; (2) "connective matter" tissue, composed of cells dispersed in an extracellular material; (3) the tissue responsible of "the specific features of each animal", in turn subdivided in nerves, vessels and blood

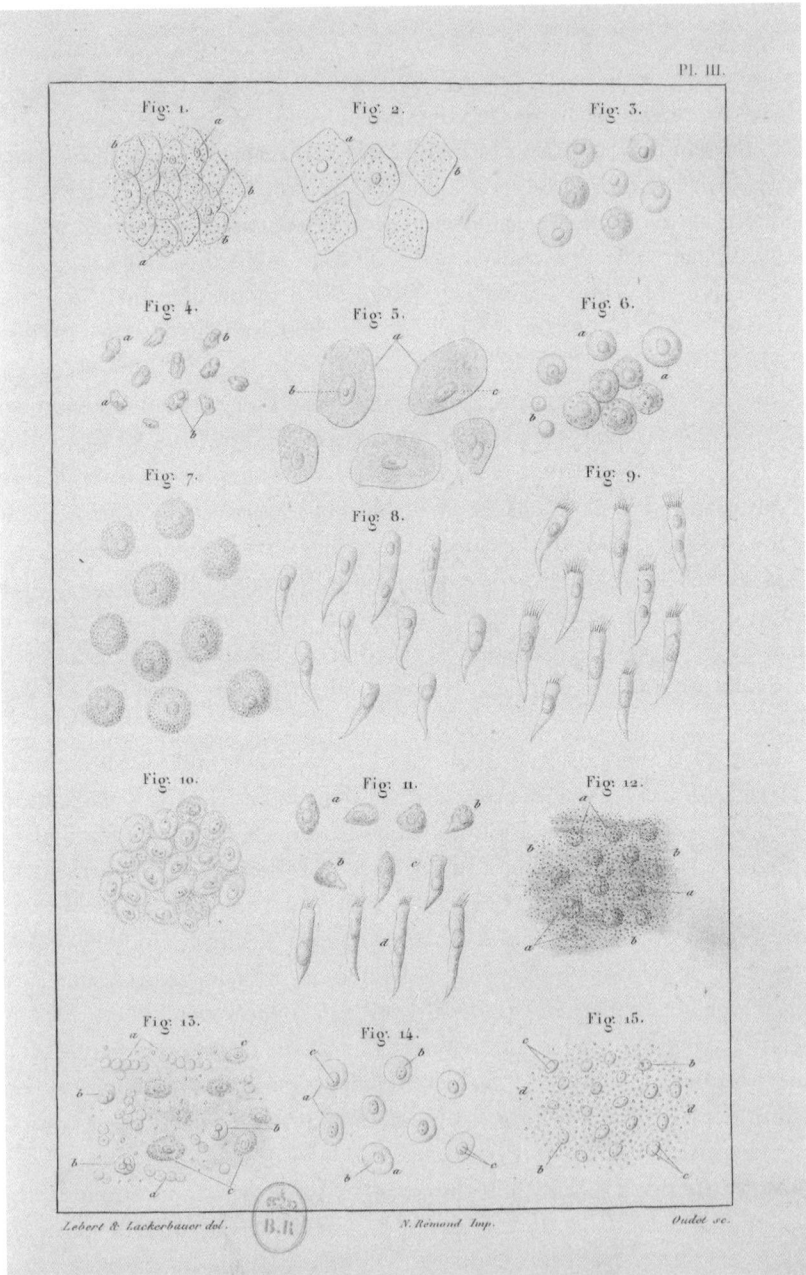

Fig. 1.2 Table III (Planche III) of Herman Lebert's Atlas (Lebert, 1845). The atlas contains 249 figures of different type of cells and lesions (visible with the naked eye, or by using a microscope), and legends are very detailed. In this table Figs. 1 and 2 show epidermal cells, Figs. 3–13, cells found in sputum. Figs. 14 and 15 show "fibro-plastic globules" (see also Fig. 1.3). *(Reproduced by kind permission of the "Bibliothèque nationale de France (BnF).")*

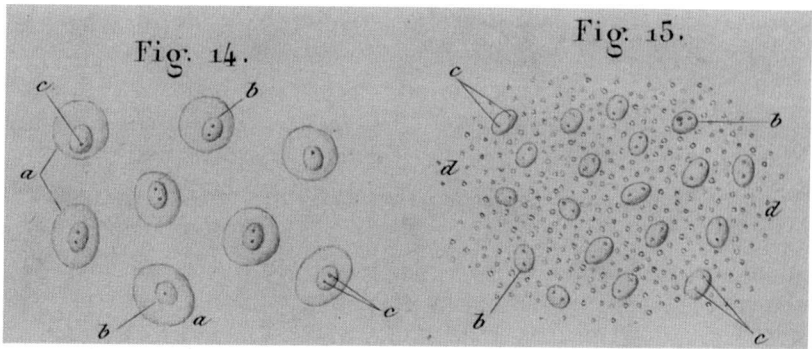

Fig. 1.3 Close-up of Lebert's Table III, showing (Fig. 14) "fibro-plastic globules" and (Fig. 15) "nucleus of fibro-plastic globules surrounded by granules" (Lebert, 1845). *(Reproduced with permission from gallica.bnf.fr/Bibliothèque nationale de France.)*

(Virchow, 1858, p. 35). Virchow considered that connective tissue was very important also because its study had been instrumental for the development of the cellular theory (Virchow, 1858, p. 44): while before connective tissue was considered to be composed only of fibers, starting from Schwann it appeared to contain also cells, e.g., the previously mentioned "fibro–cells" of Schwann or the "fibro–plastic–bodies" of Lebert.

We will consider now the controversial point of the origins of the extracellular matrix: Schwann proposed that it was derived from cell destruction originating fibrils, while nuclei remained intact. According to the German pathologist Friedrich Gustav Jakob Henle (1809–85) extracellular matrix was composed of "blastema," in which nuclei were dispersed and fibers were produced by the blastema itself. See Fig. 1.3.

Later Virchow proposed his own theory and the efforts to verify the different theories continued for the whole half of the 19th century. According to Virchow there was no transformation of cells in fibers, but, already from the beginning of the embryonic development round, fusiform and stellate cells were immersed in the fibrous extracellular matrix; in some instances round cells modulated into fusiform cells, e.g., at the surface of joints (Virchow, 1858, p. 47). See Fig. 1.4.

Altogether, Virchow tried to solve the problem of extracellular matrix origin by proposing that both cells and fibers coexisted in connective tissue from the beginning of embryonic development. For pathological tissues Virchow proposed an organization similar to that of normal tissues (Virchow, 1858, pp. 61–62). Moreover, he suggested the possibility that neoplasms could consist of a mixture of different tissues composing a sort of pathological organ (Virchow, 1858, p. 63). See Fig. 1.5.

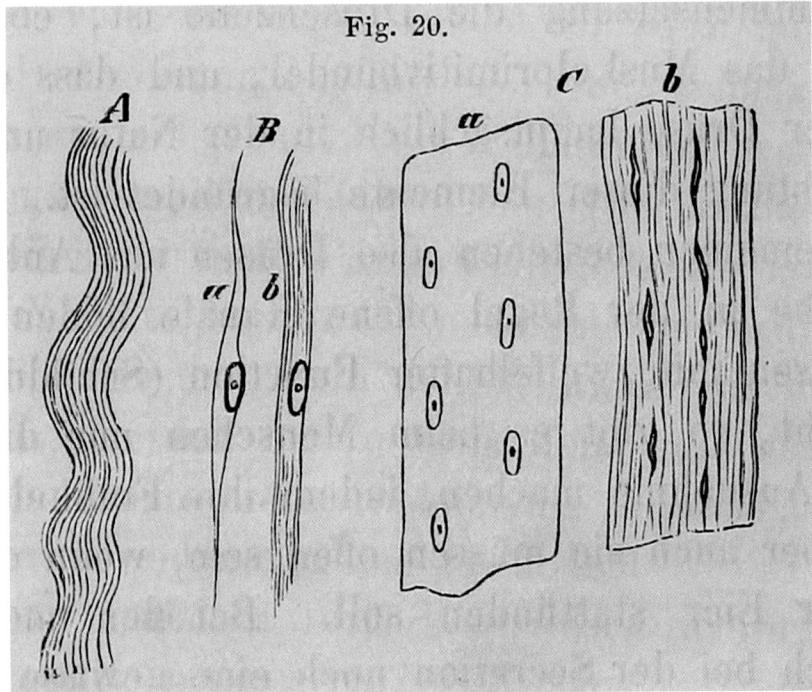

Fig. 1.4 Figure 20 from Virchow's Cellularpathologie (Virchow, 1858, p. 38). (A) Intercellular substance of connective tissue; (B) diagram of Schwann theory, according to which connective tissue is formed from the disruption of fibro-cell walls (fig. "a" represents a "fibro-plastic" "cell intact"); (C) diagram of Henle theory, according to which fibers are formed from "blastema" in which nucleus free of the cell's wall are immersed. *(From Wellcome collection.)*

In summary, three alternative views about the nature and origin of connective tissue coexisted in the second half of the 19th century (Paladino, 1887): (1) extracellular matrix was produced by the protoplasm of connective tissue cells (Schwann); (2) fibers emerged from an amorphous substance without any cellular intervention (Henle); (3) fibers were secreted by connective tissue cells, as proposed by the Italian physician Giovanni Battista Ercolani (1817–83) (Ercolani, 1866). Ercolani graduated in medicine at the University of Bologna in 1840, but his research was mainly applied to veterinary medicine; he became the director of the School of Veterinary Medicine in Turin in 1855. He was also involved in Italian Politics in the crucial period of Italy unification. In 1863 he came back to Bologna as professor of comparative anatomy in the local University where he became Dean between 1878 and 1883. His research explored several areas of animal and human anatomy and pathology, producing important contributions to the study of disease transmission from animals to man as well as to the

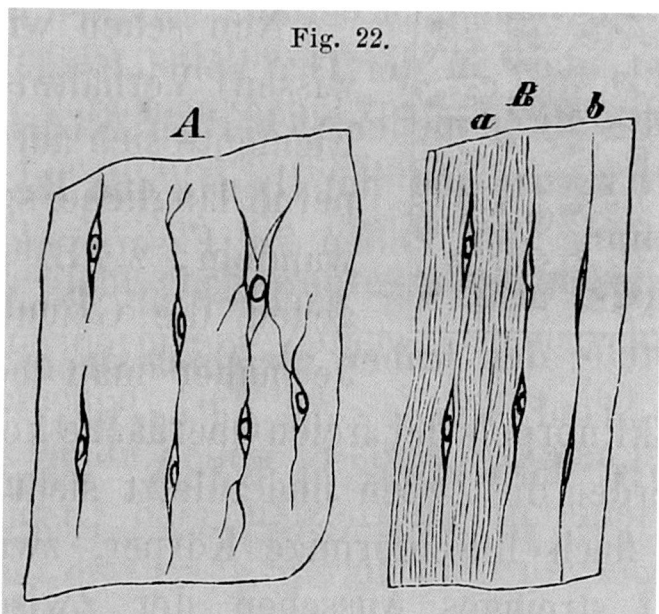

Fig. 1.5 Figure 22 from Virchow's Cellularpathologie (Virchow, 1858, p. 40). Virchow's diagram of formation of connective tissue. In the caption, Virchow indicated that the (A) represents the first stage of formation, where the cells, immerged in extracellular substance, are fusiform and divided (on the left), then becoming anastomosed and ramified (on the right). (B) The advanced stage, where: "a" shows the extracellular substance becoming striated because of the serial elongation of the cells; and "b" shows only the fibrous cells, elongated and anastomosed, while the extracellular substance is evanished because of the treatment with acetic acid. *(From Wellcome collection.)*

comparative anatomy and physiology of placenta. In the "Memoir" in 1865 entitled: "On the normal structure and the pathological alterations of the connective tissue" (original title in Italian: "Sulla Struttura Normale e sulle Alterazioni Patologiche del Tessuto Fibroso"), he states that the fibrils that occupy a great portion of connective tissue in a tendon are produced by the local elongated cells visible in between. His arguments are clearly indirect, similar to the arguments used by Schwann and Henle for their suggestions.

A significant improvement in the understanding of connective tissue structure and composition was made at the end of the 19th and the beginning of the 20th centuries thanks to several important researchers. Among them Ilya Illich Mechnikov (1845–1916), the discoverer of phagocytosis by the macrophage, who received the Nobel Prize in 1908 and was probably the most significant figure of this time. Through his work It became accepted that both "fibro-plastic" and phagocytic cells were present in connective tissue. Moreover, the German physician Paul Ehrlich (1854–1915), who also received the Nobel Prize in 1908, discovered the mast cell thanks to the use of metachromatic staining (Ehrlich, 1879). The name fibroblast was first proposed by Ernst Ziegler (1849–1905) in his book: *General Pathology: or, The science of the causes, nature, and course of the pathological disturbances which occur in the living subject* (Ziegler, 1895, p. 280; Plikus et al., 2021) in discussing the mechanisms of granulation tissue formation in an open wound and referring to a local cell primarily contributing to the production of connective tissue fibers. Finally, the most systematic investigator of connective tissue histology and biology was the Russian embryologist Alexander Maximov (1874–1928), who distinguished the following cells: fibroblast, mast cell, clasmatocyte (a term alternative to macrophage that now is no more in use), lymphocyte, plasma cell and eosinophil (Renaut, 1907). Maximov proposed that during inflammation the cells accumulating in tissues, named by him "polyblasts," originated from blood leucocytes and that these cells could eventually transform into fibroblasts (Maximov, 1927). The possibility that fibroblasts could transform into macrophages and vice-versa was also proposed later (Spadafina, 1933) and has been revived occasionally through the years up to the present time. The name fibroblast was certainly popularized by Maximov. The suffix "blast" was used to indicate a "germinative" function since the use by Schwann of the word "cytoblasteme" to design a fluid situated within the cell and able to generate a new cell through the process of "agglomeration" in a sort of nucleolus on which the other cell components were eventually condensed. The term "blast" was also used in the same period to define tumors such as "blastoma," indicating a rapid and uncontrolled growth.

In a comment appeared in the British Medical Journal on June 4, 1902 there is a discussion on the use of the term "blastoma" by well-known pathologists including Edwin Klebs (1834–1913) and George Adami (1862–1926). The article ends with this comment by Powell White:

Now, Sir, in the terms "fibroblast", "Erythroblast", etc., the termination blast signifies that the cells to which these terms are applied are immature. The term "blast" (βλαστός, germ) without any prefix would naturally signify an undifferentiated cell, that is, one which is not the precursor of any definite kind of cell, but which may become differentiated into any kind of mature cell. We have also the old term "blastemal", which was used to designate the undifferentiated substance from which the cells were supposed to be formed. The natural meaning, therefore, of the term "blastoma" is a tumour, the essential components of which are undifferentiated cells, and it is in this sense that I have used it; and I think that, both by derivation and by analogy with cognate words, I am justified in so using it.

(White, 1902)

We will not further discuss "blastema," but it appears interesting that the term fibroblast has been accepted to design an immature cell. The general assumption was that the fibroblast was the element active in the synthesis of extracellular fibers and evolved in the fixed and quiescent "fibrocyte." Unfortunately, the term "fibrocyte" has been recently used to design a bone marrow derived fibroblastic cell that invades the site of a wound or a fibrotic lesion and contributes to the formation of granulation tissue [see Chapter 2 on Wound healing, as well as Bucala, Spiegel, Chesney, Hogan, and Cerami (1994)].

In the 20th century, significant progress was achieved in the under-standing of fibroblast biology thanks to the development of new tech-niques allowing the growth of cells in culture. A pioneer in this field was the French surgeon Alexis Carrel (1873–1944), who was awarded the Nobel Prize in 1912 for his work on vascular sutures and transplanta-tion. His work in collaboration with Albert H. Ebeling described in de-tail many features of cultured fibroblasts and their biological differences with macrophages (Carrel & Ebeling, 1926). Carrel produced also an important contribution on the biology of wound healing (as we shall see Chapter 2) and on the formation and contraction of granulation tissue. His work furnished the basis for the hypothesis that the fibroblast was crucial for wound contraction.

We think that at this point ends the history and start the present time. It appears that in the first quarter of the 20th century the role of the fibroblast in the building and organization of extracellular matrix was well accepted and that his role in wound healing and granulation tissue formation was seriously considered.

References

Benedetti, A. (1502). *Anatomice sive historia corporis humani*. Venezia: Bernardino Guerralda vercellese.

Berg, A. (1942). Die Lehre von der Faser als Form- und Funktionselement des Organismus. *Virchows Archiv für Pathologische Anatomie und Physiologie und für Klinische Medizin, 309*(2), 333–460. https://doi.org/10.1007/bf02593519.

Bichat, M. F. X. (1812). *Anatomie générale, appliquée à la physiologie et à la médecine*. Paris: Brosson, Gabon et Cie.

Bucala, R., Spiegel, L. A., Chesney, J., Hogan, M., & Cerami, A. (1994). Circulating fibrocytes define a new leukocyte subpopulation that mediates tissue repair. *Molecular Medicine, 1*(1), 71–81.

Carrel, A., & Ebeling, A. H. (1926). The fundamental properties of the fibroblast and the mac-rophage: I. The fibroblast. *The Journal of Experimental Medicine, 44*(2), 261–284. https://doi.org/10.1084/jem.44.2.261.

Ehrlich, P. (1879). Beiträge zur Kenntnis der granulierten Bindegewebszellen und der eosin-ophilen Leucocyten. *Archives of Anatomy and Physiology*, *3*, 166.

Ercolani, G. B. (1866). *Osservazioni sulla struttura normale e sulle alterazioni patologiche del tessuto fibroso*. Bologna: Gamberini e Parmeggiani.

Falloppia, G. (1575). *Lectiones Gabrielis Fallopii De partibus similaribus humani corporis, ex diversis exemplaribus a Volchero Coiter summa cum diligentia collectae*. Norimbergae: In officina Theodorici Gerlachi.

Ferrari, G. (1996). *L'esperienza del passato: Alessandro Benedetti filologo e medico umanista* (pp. 119–120). Florence: Leo S. Olschki.

Hajdu, S. I. (2004). The first cellular pathologists. *Annals of Clinical and Laboratory Science*, *34*(4), 481–483.

Hunter, J. (1794). *A treatise on the blood, inflammation, and gun-shot wounds*. London, UK: J. Richardson for G. Nicol.

Hyrtl, J. (1884). *Onomatologia anatomica. Storia e critica del moderno linguaggio anatomico*. Roma: Voghera Carlo, Tipografo di S. M.

Kaiser, H. (2010). Hermann Lebert (1813–1878). *Zeitschrift für Rheumatologie*, *69*(5), 461–468. https://doi.org/10.1007/s00393-010-0615-3.

Keill, J. (1698). *The anatomy of the human body, abridged: Or, a short and full view of all the parts of the body. Together with their several uses, drawn from their compositions and structures*. London: William Keblewhite.

Lebert, H. (1845). *Physiologie pathologique ou recherches cliniques, expérimentales et microscopiques sur l'inflammation, la tuberculisation, les tumeurs, la formation du cal, etc. Atlas de vingt-deux planches*. Paris: J.-B. Ballière.

Marcovecchio, E. (1993). *Dizionario etimologico storico dei termini medici* (p. 1993). Impruneta-Firenze: Festina Lente.

Maulitz, R. (1987). *The morbid appearances. The anatomy of pathology in the early nineteenth century*. New York: Cambridge University Press.

Maximov, A. (1927). Development of non-granular leucocytes (lymphocytes and monocytes) into polyblasts (macrophages) and fibroblasts in vitro. *Experimental Biology and Medicine*, *24*, 570–572.

Meckel, J. F. (1838). *Manual of general, descriptive, and pathological anatomy*. Vol. I. Philadelphia, PA: Carey & Lea. 1832.

Ongaro, G. (1981). La medicina nello Studio di Padova e nel Veneto. In M. P. Stocchi, & G. Arnaldi (Eds.), *Vol. 3/III. Storia della cultura veneta. Dal primo Quattrocento al Concilio di Trento* (pp. 75–134). Vicenza: Neri Pozza Editore.

Paladino, G. (1887). *Ulteriori ricerche sulla distruzione e rinnovamento continuo del parenchima ovarico nei Mammiferi: nuove contribuzioni alla morfologia e fisiologia dell'ovaja*. Napoli: Antonio Morano.

Pickel, H., Reich, O., Winter, R., & Young, R. H. (2009). Hermann Lebert (1813–1878): A pioneer of diagnostic pathology. *Virchows Archiv*, *455*(3), 301–305. https://doi.org/10.1007/s00428-009-0820-0.

Plikus, M. V., Wang, X., Sinha, S., Forte, E., Thompson, S. M., Herzog, E. L., … Horsley, V. (2021). Fibroblasts: Origins, definitions, and functions in health and disease. *Cell*, *184*, 3852–3872.

Renaut, J. (1907). Les cellules connectives rhagiocrines. *Archives d'anatomie microscopique*, *IX*, 495–609.

Robb-Smith, A. H. T. (1954). Normal morphology and morphogenesis of connective tissue. In G. Asboe-Hansen (Ed.), *Connective tissue in health and disease*. Copenhagen: Ejnar Munksgaards Forlag.

Schwann, T. (1839). *Mikroskopische Untersuchungen über die Uebereinstimmung in der Struktur und dem Wachsthum der Thiere und Pflanzen*. Berlin: Sander.

Skinner, H. (1970). *The origin of medical terms*. New York, NY: Hafner Publishing Company.

Spadafina, L. (1933). Intorno al problema della trasformazione dei fibroblasti in macrofagi. Ricerche sperimentali. *Zeitschrift für Zellforschung und Mikroskopische Anatomie, 18*(1), 192–216. https://doi.org/10.1007/BF01094515.

Vesalius, A. (1543). *De humani corporis fabrica libri septem.* Basilea: Officina Ioannis Oporini.

Virchow, R. (1858). *Die cellularpathologie in ihrer begründung auf physiologische und pathologische gewebelehre. Zwanzig vorlesungen gehalten während der monate februar, märz und april 1858 im Pathologischen institute zu Berlin.* Berlin: A. Hirschwald.

Vogel, J. (1843). *Histologiae pathologicae. Tabulae histologiam pathologicam illustrantes. Vigini sex tabulae, continents CCXCI figuras, quarum CCLXX ad naturam delineatae sunt.* Lipsiae: Leopoldus Voss.

von Haller, A. (1757). *Elementa physiologiae corporis humani. Vol. 1.* Lausannae: Sumptibus Marci-Michael Bousquet & Sociorum.

White, C. (1902). On the use of the term "blastoma". *The Lancet, 159*(4111), 1723–1724. https://doi.org/10.1016/S0140-6736(01)85632-4.

Zampieri, F. (2016). *Il metodo anatomo-clinico tra meccanicismo ed empirismo: Marcello Malpighi, Antonio Maria Valsalva, Giovanni Battista Morgagni.* Roma: "L'Erma" di Bretschneider.

Ziegler, E. (1895). *General Pathology: or, The science of causes, nature and course of the pathological disturbances wich occur in the living subject* (p. 280). Wood and Company.

CHAPTER 2

Wound healing and wound contraction

In the first half of the 20th century, the concept that wound healing was accomplished through two main processes gradually emerged: (1) epithelialization, i.e., replication and movement of epithelial cells and (2) formation and contraction of granulation tissue. The observation that granulation tissue fibroblasts are equipped with a contractile apparatus opened the way to the concept of the myofibroblast.

Tissue repair and wound healing is one of the longest standing and central subjects in medicine and surgery (Abdelaal, Giovinco, Slepian, & Armstrong, 2015, p. 563). The early history of wound healing is related to the "clinical practice" rather to the understanding of physiological mechanisms. These ancient practices where based more on empiricism rather than on physiological and pathophysiological theories (Majno, 1975).

A Mesopotamian clay tablet (the so-called "Medical Tablet," or tablet B14221; Fig. 2.1) dating to 2500–2340 BCE and found during the "Babylonian expedition" (1888–1900) of the University of Pennsylvania to the ancient Sumerian city of Nippur, Iraq, documents a multistep approach to wound care management. This tablet, indeed the oldest medical written source, is inscribed with 15 medical prescriptions. Eight of these prescriptions (but only prescriptions 4–8 are enough conserved to be correctly interpreted) describe wound treatment by application of topical remedies (poultices) to a wound. Each prescription comprises four moments [commonly referred as to the "three healing gestures" since Majno's (1975) "The Healing Hand," see also Shah (2011)]: 1. preparation of the ingredients by pulverization, 2. pouring beer, water-diluted beer, or water tout court on the ingredients to make a knead, 3. rubbing beforehand of the wound with oil, and 4. fasten as a poultice (e.g., prescription 4: "[1] Pulverize the anadishsha-plant, the branches of the" thorn'-plant, the seeds of the duashbur…[2] pour water-diluted beer over it… [3] rub with vegetable oil… [4] fasten the paste"; prescription 7: "[1] Pulverize the lees of the dried vine,

17

Fig. 2.1 Penn Museum, object no. B14221. The Mesopotamian clay tablet B14221, or "Medical Tablet" (2500–2340 BCE), represents the oldest medical text. It contains 15 medical prescriptions; 8 of these are dedicated to wound treatment. *(Reproduced with permission from Penn Museum. All rights reserved.)*

pine tree, and plum tree… [2] pour beer over it… [3] rub with oil… [4] fasten as a poultice)" (Kramer, 1963).

In the 20 m long Egyptian "Ebers Papyrus" (~1500 BC),[a] it is outlined the use of lint, animal grease, and honey (still used in wound dressings) as topical treatment of wounds (Abdelaal et al., 2015). Although the Ebers papyrus is the largest record of ancient Egyptian medicine, the smaller (4.7 m long)

[a] The "Ebers papyrus" was composed around 1300 BC. It was purchased by the German (Leipzig) explorer and Egyptologist Georg Ebers (1837–98) from Edwin Smith (see note 4) in 1872, in Thebes. Still, it is unclear from whom Smith purchased this papyrus. It is actually conserved at the University Library of Leipzig. It represents the longest and more complete Egyptian medical text, and it contains a practical and theoretical knowledge dating back to a period much older than 1300 BC.

"Smith papyrus"[b] (or Edwin Smith Surgical Papyrus, ~1500 BC), is indeed a "book of wounds." It presents 48 cases studies of traumatic injuries to the head and torso (including spinal cord injuries), and focuses on diagnosis (probing with the fingers) and treatment of wounds (Hartmann, 2016). Treatment included both wound, as well as sutures, reducing of fractures, splints and 8 spells (Moore, 2011). As an example, case number 10, "Instructions concerning a wound in his head penetrating to the bone of his skull," precisely describes how to take care of the surgical wound (after drawing "together for him the gash with stitching"):

Now after thou hast stitched it, thou shouldst bind fresh meat upon it the first day. If thou findest that the stitching of this wound is loose, thou shouldst draw (it) together for him with two strips (of plaster), and thou shouldst treat it with grease and honey every day until he recovers.

(Breasted, 1930, pp. 225–233)

Until the 17th century, Western medicine was dominated by Hippocrates's theory of the "humors," which stated that human physiology depends upon the balance of the four humors, i.e. "blood," "phlegm," "black bile" and "yellow bile." Each one of these humors was related to one of the four "elements" and a couple of the four "qualities" of classic physics, so that blood was "heat" and "humid" and related to "air," phlegm was "cold" and "humid" and related to "water," black bile was "cold" and "dry" and related to "earth," and yellow bile was "heat" and "dry" and related to "fire" (Arikha, 2007).

Although the ancients observed and described blood coagulation and understood that it was a fundamental process in wound healing, this phenomenon, alike scab and scar formation, was attributed to phlegm and then to black bile (because of its coldness, and dryness); only the accompanying inflammation was attributed to blood. The Roman encyclopedist and physician Aulus Cornelius Celsus (14–73 CE) first described the four cardinal signs of inflammation: *calor* (warmth), *dolor* (pain), *tumor* (swelling) and *rubor* (redness and hyperemia)[c]

---

[b] The "Smith papyrus" was composed around 1600 BC, but is a partial copy of an earlier document (3000–2500 BCE) (Wilkins, 1964). Like the Ebers Papyrus, the Smith Papyrus was purchased by the American Edwin Smith (1822–1906), one of the earliest students of Egyptian hieroglyphics. However, the name of the dealer from whom he bought the papyrus in Luxor is known: Mustapha Aga. After Smith's death (1906), his daughter Leonora donated the papyrus to the New York Historical Society, which entrusted the papyrus to the founder of the Oriental Institute in Chicago, James H. Breasted, with the task of translating it in 1920.

[c] The original Celsus' aphorism is: "*Notae vero inflammationis sunt quattuor: rubor et tumor cum calor et dolor.*" ("Now the signs of an inflammation are four: redness and swelling with heat and pain.") It can be found in book 1 of his *De Medicina* (On Medicine); see Spencer's 1935 translation (Celsus, 1935).

(Mitchinson, 1989). These terms became widely used in medical language, and still are nowadays. Moreover, these signs could be perfectly explained by the supposed qualities of blood. It is commonly believed that it was Galen, a century and a half later, that added to Celsus tetrad the famous fifth cardinal sign of inflammation: *functio laesa* (disturbance of function). However, the paternity of this term, as well as the original Greek term, is disputed (Rather, 1971).

However, in the history of surgery, we can find some interesting observation also during the period in which humors dominated medicine. One of the most important surgeons of late Middle Age, for instance, was Bruno da Longobucco. Born in south Italy, around the beginning of 1200, he became professor at the University of Padua, where he died in 1286. Around the 1250s, he composed his *Chirurgia Magna*,[d] in which he was one of the firsts to describe wound healing by first and second intention (Fig. 2.2). In the first chapter of this work, he writes[e]:

> I say that the wound can be simple or complex [...]. Simple is when the separation of organic matter can be cured simply through the union of detached parts; complex is when the solution of continuity involves the loss of organic matter. In this case, there are two types of cure: I) the union of detached parts; II) the regeneration of lost organic matter.

> **(Longobucco, 1499, p. 5; see also Focà, 2004, p. 71)**

In the second chapter, Longobucco is even more precise[f]:

> [...] it is possible to restore the solution of continuity in the flesh by applying the first intention of cure, while the solution of continuity in the bones follows only the second intention. The first intention is the consolidation, which cannot be done in the bones because of their hardness. The second intention is the conjunction of the margins, which is made possible by the formation of sarcoid tissue [...].

> **(Longobucco, 1499, p. 16)**

---

[d] The work was published for the first time in Venice in 1498, well after Longobucco's death, in a book containing different surgical treatises, entitled *Cyrurgia Guidonis de Cauliaco et Cyrurgia Bruni Theodorici Rogerii Rolandi Bertapalie Lanfranci*.

[e] In Italian, in the original text: "dico, dunque, che la soluzione di continuo (ferita), malattia comune, si divide in semplice e composta[...] Semplice è quella nella quale vi è solo il taglio di sostanza e ha un solo modo di cura, cioè la unione delle parti staccate; composta è la soluzione di continuo nella quale esiste perdita di sostanza. Vi sono due tipi di cure: I) l'unione delle parti staccate; II) la rigenerazione della sostanza perduta."

[f] In Italian, in the original text: "[...] la soluzione di continuità nella carne è possibile che ritorni così come è stata seguendo la prima intenzione di cura, invece la soluzione di continuità nelle ossa non si guarisce attraverso di essa, ma soltanto con la seconda intenzione. La prima intenzione è quella della consolidazione, che non può invece avvenire nelle ossa a causa della loro durezza. La seconda intenzione è la saldatura, che deve avvenire con la formazione di un polo sarcoide [...]."

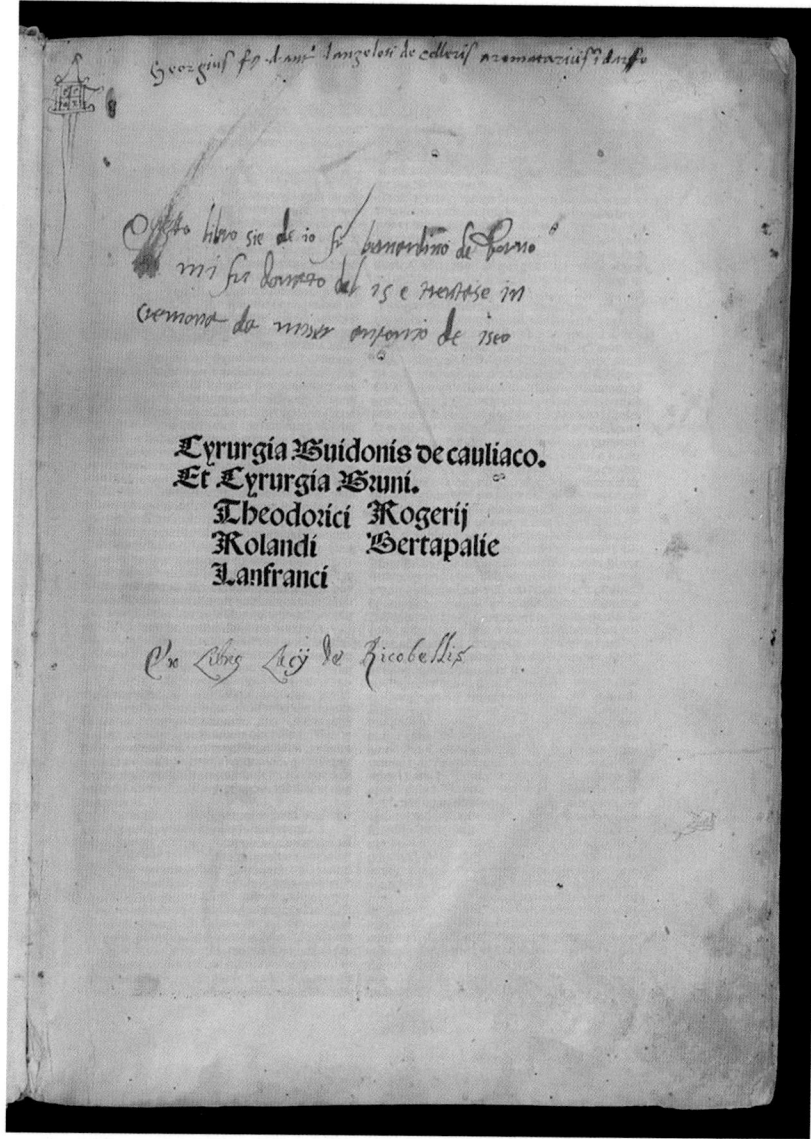

Fig. 2.2 Frontispiece of the first edition of the Longobucco "Chirurgia Magna" (1498), inside a collective work of surgery written by Longobucco, Guy de Chauliac (about 1300–68) and Leonard of Bertipaglia (about 1380–1463). Hence the title: *Cyrurgia Guidonis de cauliaco* (i.e., Guy de Chauliac). *Et Cyrurgia Bruni (scil. Longoburgensi*; i.e., Bruno da Longobucco) *Theodorici Rolandi Lanfranci Rogerii Bertapalie* (Leonard of Bertipaglia). Edition of the early 16th century, from the library of Walter Pagel (1898–1983; German pathologist and medical historian). The *Chirurgia Magna* was finished at Padua, Italy, where Longobucco was professor of surgery, it became one of the most important surgical texts in European universities for almost two centuries. It was published for the first time in 1498. To note: the marginal annotations say: "1. Georgius fn. d.mi lanzeloti de celleris aromatarius i[n] darffo. 2. Questo libro sie de io fr. bernardino de Corno [et ?] mi fu donato del 15 e trentase [i.e. 1536] in cremona da miser antonio de iseo; 3. Ex libris Luey de Ricobellis." *(Reproduced with permission from Wellcome Collection. Attribution 4.0 International (CC BY 4.0).)*

Only at the beginning of the modern era, from the XVII century, the theory of humors was seriously questioned and new ideas about wound healing started to be proposed and developed. The British Surgeon John Hunter (1728–1793), with his famous *Treatise on the Blood, Inflammation and Gun-shot Wound*, published posthumously (Hunter, 1794), marked a fundamental advancement in the knowledge of wound healing; his pioneering ideas, were not fully understood and proved until the XIX and XX centuries (Turk, 1994). Hunter reported many interesting observations stemmed from his clinical experience (in particular his practice as a military surgeon in the Seven Years' War [1756–1763]), as well as from his experimental research with humans and animals. He was able to demonstrate, for instance, that the scarlet color of arterial blood is produced by the contact with air through the lungs, and that blood is composed of "globules" (red cells) and coagulable lymph or serum. Moreover, he described elastic and muscular features of arteries, showed experimentally that contraction is responsible for arterial lumen reduction and that small arteries are more contractile than large ones, such as the aorta.

One of Hunter's fundamental achievements was to understand the importance of granulation tissue formation (called by him "granulating or incarnation and the substance formed is called granulations") in the repair process and in particular, that granulating is the consequence of inflammation: "[…] inflammation is in general necessary […] for disposing the vessels to form granulations." He also states that granulations are independent of suppuration. The description of "granulations" is very clear: "Their surfaces are very convex, the reverse of ulcerations, having a great many points, or small eminences, so as to appear rough: and the smaller these points are, the healthier we find the granulations." Hunter gave also detailed description of what he called "contraction of granulations": "The contraction takes place in every point, but principally from edge to edge, which brings the circumference of the sore towards the center, so that the sore becomes smaller and smaller, although there is little or no new skin formed." Interestingly, Hunter notes that contraction is more important in "long" compared to round sores.

Thanks to experiments on the rabbit's ear and cock's wattle, Hunter was able to demonstrate first that "inflammation" is the fundamental process of the returning of "life" to a part of the body after freezing. Then, he proved that blood is involved in tissue repair by extravasation and coagulation as well as by the formation in the coagulum of new blood vessels, which infiltrate from the surrounding areas and form the basis for the development of granulation tissue. The increasing number of small vessels appeared particularly evident in the inflammation following freezing.

According to Hunter's observations in gunshot wounds, a kind of injury that communicates externally from the body, the blood clot becomes dry and forms a scab, producing a union of the wound by "first intention." Where union by first intention does not take place, "adhesive inflammation" occurs. Sometimes, and particularly in the presence of a foreign body, a "suppurative inflammation" can occur before the "adhesive inflammation." If adhesive inflammation does not occur, a third form of union arises and this is by "granulations." Hunter noted that the possibility of union can follow also transplantation of tissue. The spur of a cock, for instance, can grow on its comb. Hunter noted that this process is similar to the first intention healing of wounds. A similar idea was developed a century later by the German surgeon Karl Thiersch (1822–1895). Thiersch attempted to grow the skin cells into granulating wounds. In doing this, he confirmed the role of granulation tissue on wound healing (Thiersch, 1874).

Hunter understood that the formation of the scab is an important part of the healing, that inflammation occurs around the scab and suppuration below it. He also observed that inflammation could be reduced by cold, but that this could not be used as therapy "because the process is reparatory."

Between the XIX and the XX century, the study of wound healing improved substantially thanks to two theoretical revolutions (the cell theory and the immune system) and one technical advancement (cell culture). The development of the cell theory focused the researchers' attention to the different cell categories involved in tissue repair. In the same period the existence of the immune system was postulated on the basis of the discovery of phagocytosis. Cell culture, allowing the growing of cells in vitro, permitted to study several aspects of cell differentiation and proliferation in wound healing.

The German pathologist Rudolph Virchow (1821–1902), considered the founder of cell pathology, proved that cells arise only from preexisting cells (*omnis cellula ex cellula*), thus confuting Matthias Jacob Schleiden's (1804–81) and Theodor Schwann's (1810–82) ideas of spontaneous cell generation. In the context of wound healing, Virchow supported the assumption that tissue regeneration depends on cell proliferation, thus opening the way to the investigation of the cells involved in this process. A notable contribution of Virchow was the pathophysiological theory of atherosclerosis (*endarteritis deformans*, in his own words), based on the description of several atherosclerotic lesions at different stages. The "endarteritis" was seen as the product of an inflammatory process within the intima, initiated by mechanical forces, whereas the fibrous thickening was the consequence of a reactive fibrosis and part of a repair mechanism and thus confirmed that inflammation was the basic process in tissue repair. Virchow's intuition that intimal injury is the initiating irritating stimulus is still accepted at present (Mayerl et al., 2006).

Another fundamental theoretical advancement for the understanding of wound healing came from the discoveries of Ilya Metchnikoff, who theorized that the purpose of inflammation was to bring phagocytic cells to the injured area to engulf bacteria. Following this theory, the Austrian Pathologist Franz Josef Lang (1894–1975) published in 1926 that macrophages arise from local precursors and circulating monocytes. He also suggested that injured endothelial cells will elongate and differentiate into wound "fibroblasts" (Lang, 1926). The American physician and experimental pathologist Leo Loeb (1869–1959) was the first to attempt to grow cells outside the human body (Loeb, 1897). However, he could only obtain their survival, but not their growth (van Winterswijk & Nout, 2007). The milestone breakthrough in in vitro cell culture was achieved by the American biologist Ross Granville Harrison (1870–1959), who demonstrated the active growth of cells in culture (Harrison, 1910). In particular, he was the first to grow frog ectodermal cells in vitro, thus developing the first neuronal tissue culture line. Since that time, cell biology and particularly in vitro cell culture became the mainstay of what can be considered classical tissue engineering (Meyer, 2009). Harrison was followed in 1912 by the French surgeon Alexis Carrel (1873–1944) who was able to grow pieces of chick embryo in various media: he initially maintained them for 85 days, and subsequently for years (Carrel, 1912; van Winterswijk & Nout, 2007). Later, interest arose in growing cells instead of complete tissues, in particular following the discovery that trypsin is capable of degrading matrix proteins and thus of separating cells (Rous & Jones, 1916; van Winterswijk & Nout, 2007). In the first half of the 19th century gradually emerged the concept that wound healing was accomplished through two main processes: (1) epithelialization i.e. replication and movement of epithelial cells and (2) formation and contraction of granulation tissue. This last phenomenon, allegedly due to forces generated within the tissue itself (Carrel, 1916), depended, according to two antithetic theories, either to collagen fiber rearrangement and or contraction or to fibroblast contraction. The first theory, more traditional, was summarized by G. Payling Wright in the textbook "Introduction to Pathology" (Payling Wright, 1954) and was supported, among others, by J. Gros, well known for the discovery of collagenase, who suggested that collagen crosslinking was an important player in this phenomenon. In addition, Majno (1958) had shown that heating collagenous tissues, such as tendon, results in their retraction and suggested that this non physiological behavior could help in explaining tissue contraction in vivo (Majno, 1958). The second theory was supported by M. Abercrombie, who had shown that scorbutic guinea pigs do not exhibit wound contraction during the healing of an experimental open wound (Abercrombie, Flint, & James, 1956) and H. Hoffman-Berling, who had shown that glycerinated cultured fibroblasts contract upon adenosine triphosphate treatment (Hoffman-Berling, 1954). As we shall see below, the observation that granulation tissue fibroblasts are equipped with a contractile apparatus opened the way to the concept of the myofibroblast (Gabbiani, Ryan, & Majno, 1971).

References

Abdelaal, M. G., Giovinco, N. A., Slepian, M., & Armstrong, D. (2015). Tissue repair and wound healing: A trip back to the future. In *Technological advances in surgery, trauma and critical care* (pp. 563–571). New York: Springer.

Abercrombie, M., Flint, M. H., & James, J. W. (1956). Wound contraction in relation to collagen formation in sorbitic guinea pigs. *Journal of Embryology and Experimental Morphology, 4*, 167–180.

Arikha, N. (2007). *Passions and tempers: A history of the humours.* New York: HarperCollins.

Breasted, J. (1930). *The Edwin smith surgical papyrus.* Chicago, IL: University of Chicago Press.

Carrel, A. (1912). On the permanent life of tissues outside of the organism. *The Journal of Experimental Medicine, 15*, 516–528.

Carrel, A. (1916). Cicatrisation of wounds. I. The relation between the size of a wound and the rate of its cicatrisation. *The Journal of Experimental Medicine, 24*, 429–550.

Celsus, A. (1935). *On medicine, volume I: Books 1–4* (W. Spencer, Trans.). Cambridge, MA: Harvard University Press.

Focà, A. (2004). *Maestro Bruno da Longobucco chirurgo.* Reggio Calabria: Laruffa Editore.

Gabbiani, G., Ryan, G. B., & Majno, G. (1971). Presence of modified fibroblasts in granulation tissue and their possible role in wound contraction. *Experientia, 27*(5), 549–550. https://doi.org/10.1007/bf02147594.

Harrison, R. G. (1910). The outgrowth of the nerve fiber as a mode of protoplasmic movement. *Journal of Experimental Zoology, 9*(4), 787–846.

Hartmann, A. (2016). Back to the roots—Dermatology in ancient Egyptian medicine. *Journal der Deutschen Dermatologischen Gesellschaft, 14*(4), 389–396.

Kramer, S. N. (1963). *The Sumerians. Their history, culture, and character.* Chicago: University of Chicago Press.

Hoffman-Berling, H. (1954). Adenosintriphosphat als Betriebsstoff von Zellbewegungen. *Biochimica et Biophysica Acta, 14*, 182–195.

Hunter, J. (1794). *A treatise on the blood, inflammation and gun-shot wounds.* London: Nicol.

Lang, F. J. (1926). Role of endothelium in the production of polyblasts (mononuclear wandering cells) in inflammation. *Archives of Pathology & Laboratory Medicine, i*, 41.

Loeb, L. (1897). *Ueber die Entstehung von Bindegewebe, Leukozyten und roten Blutkörperchen aus Epithel und über eine Methode, isolierte Gewebsteile zu züchten.* Chicago: M. Stern und Co.

Longobucco, B. (1499). *Cyrurgia Guidonis de Cauliaco. Et Cyrurgia Bruni Theodorici Rogerii Rolandi Bertapalie Lanfranci*. Venice: impensis Andreae Torresani de Asula.

Majno, G. (1958). Contraction of collagen fibres in vivo induced by inflammation. *Lancet, 2*, 994–996.

Majno, G. (1975). *The healing hand. Man and wound in the ancient world* (2012/08/16 ed.). Cambridge, MA: Cambridge University Press.

Mayerl, C., Lukasser, M., Sedivy, R., Niederegger, H., Seiler, R., & Wick, G. (2006). Atherosclerosis research from past to present—On the track of two pathologists with opposing views, Carl von Rokitansky and Rudolf Virchow. *Virchows Archiv, 449*(1), 96–103. https://doi.org/10.1007/s00428-006-0176-7.

Meyer, U. (2009). The history of tissue engineering and regenerative medicine in perspective. In U. Meyer, J. Handschel, H. P. Wiesmann, & T. Meyer (Eds.), *Fundamentals of tissue engineering and regenerative medicine*. Berlin, Heidelberg: Springer.

Mitchinson, M. J. (1989). Fluor: Another cardinal sign of inflammation. *Lancet, 2*(8678–8679), 1520. https://doi.org/10.1016/s0140-6736(89)92957-7.

Moore, W. (2011). The Edwin Smith papyrus. *BMJ, 342*, d1598. https://doi.org/10.1136/bmj.d1598.

Payling Wright, G. (1954). *An introduction to pathology*. London: Longmans Green & Co.

Rather, L. J. (1971). Disturbance of function (functio laesa): The legendary fifth cardinal sign of inflammation, added by Galen to the four cardinal signs of Celsus. *Bulletin of the New York Academy of Medicine, 47*(3), 303–322.

Rous, P., & Jones, F. S. (1916). A method for obtaining suspensions of living cells from the fixed tissues, and for the plating out of individual cells. *Journal of Experimental Medicine, 23*(4), 549–555.

Shah, J. B. (2011). The history of wound care. *The Journal of the American College of Certified Wound Specialists, 3*(3), 65–66. https://doi.org/10.1016/j.jcws.2012.04.002.

Thiersch, C. (1874). Über die feineren anatomischen Veränderungen bei Aufheilung von Haut auf Granulationen. *Langenbecks Archiv für Chirurgie, 17*, 318.

Turk, J. L. (1994). Inflammation: John Hunter's "A treatise on the blood, inflammation and gun-shot wounds". *International Journal of Experimental Pathology, 75*, 385–395.

van Winterswijk, P. J., & Nout, E. (2007). Tissue engineering and wound healing: An overview of the past, present, and future. *Wounds, 19*(10), 277–284.

Wilkins, R. H. (1964). Neurosurgical Classic—XVII. *Journal of Neurosurgery, 21*(3), 240. https://doi.org/10.3171/jns.1964.21.3.0240.

CHAPTER 3

Fibrotic diseases

3.1 Hypertrophic scars and keloids

The earliest recognizable description of a keloid can be found in the third enlarged edition of the *"Diseases of the Skin"* (*Des maladies de la peau*) published by the French physician Noël Retz (1758–1810) in 1790. Here, he dedicated a paragraph describing what he called a fatty pustule (*"dartre graisseuse"*), which can be recognized as a keloid (Retz, (1790), pp. 155–158; Petit, 2016, p. 81). The first to coin this specific term was the French dermatologist Jean-Louis-Marc Alibert (1768–1837), using the French *"chéloïde"* instead of the previously used term *"cancroïde"* for avoiding possible confusion between cancer and this kind of specific lesion (Alibert, 1810; Petit, 2016) In his *Précis théorique et pratique sur les maladies de la peau*, Alibert also published the first illustration of a keloid (Alibert, 1810, p. 28; Petit, 2016) See Fig. 3.1. In the course of the 19th century, following the first description of Alibert, the keloid lesion was recognized as an independent clinical entity, while the etiology remained under dispute. The most common clinical signs mentioned by the dermatologists were itch and pain. At the end of the century, the Hungarian physician and dermatologist Moritz Kaposi (1837–1902) gave a first microscopical description of keloid lesion, noting *"thick bundles of fibers compressing blood vessels"* (Kaposi, 1881; Petit, 2016).

Hypertrophic scars and keloids are long known complications of wound healing (Ghazawi, Zargham, Gilardino, Sasseville, & Jafarian, 2018). It is not always easy to distinguish between the two lesions: Hypertrophic scars do not exceed the wound margins while keloids tend to go beyond wound limits, sometime very extensively. Histologically, hypertrophic scars contain regularly fibroblastic nodules that express large amounts of α-SMA, while fibroblastic cells in keloids show a low expression of this protein (Ghazawi et al., 2018). In hypertrophic scars collagen bundles are relatively regular while keloids show large hyalinized bundles. Hypertrophic scars regress generally with time and tend to produce contractures, while keloids do not. The high expression of α-SMA in myofibroblasts of hypertrophic scars

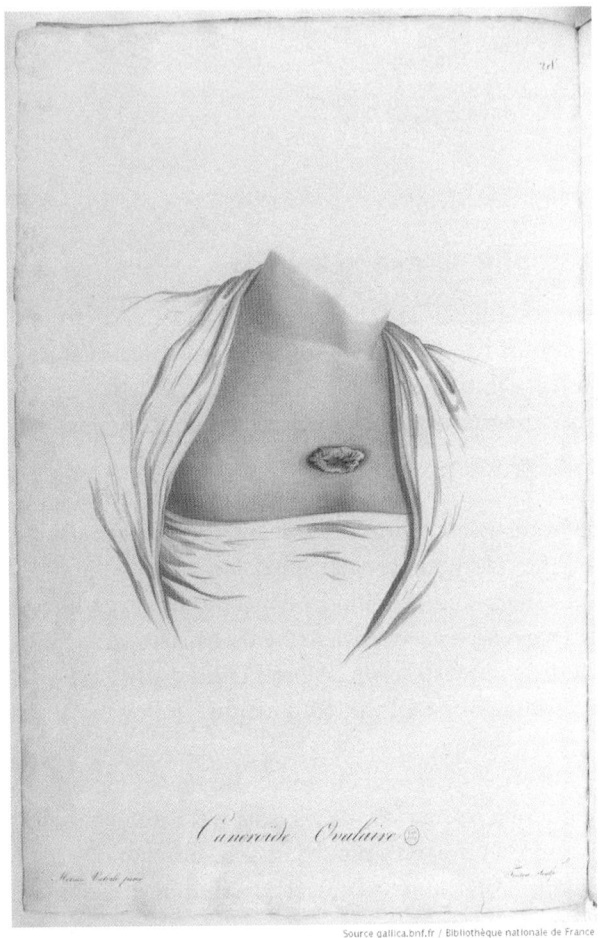

Fig. 3.1 The illustration shows an oval chancroid (*cancroid ovalaire*), later keloid, from Alibert's « *Précis théorique et pratique sur les maladies de la peau* » (p. 28). (Source: gallica. bnf.fr/Bibliothèque nationale de France.)

correlates well with their activity in producing contractures (Ehrlich et al., 1994; Santucci, Borgognoni, Reali, & Gabbiani, 2001).

3.2 Fibromatoses and Dupuytren disease

Fibromatoses are a group of diseases characterized by the accumulation of fibroblastic cell, mainly myofibroblasts, in various connective tissue areas with, in many instances, retraction phenomena in surrounding tissues and, often, functional impairment (Zhang & Kargel, 2018). They may be located superficially or deeply extra-abdominally or intra-abdominally. The most studied are those affecting the palmar and plantar aponeuroses (whose eponyms are Dupuytren and Ledderhose disease, respectively), Lapeyronie disease (*induratio penis plastica*) a fibromatous process interesting the *tunica albuginea* of the *corpora cavernosa* of the penis, and the knuckle pads small, peri-articular, subdermal nodules (also called Garrod pads, holoderma, interphalangeal pads, *keratosis supracapitularis, pulvinus*, subcutaneous fibroma (Ronchese, 1966).

What is now called as "Dupuytren disease" was thoroughly described by the French physician baron Guillaume Dupuytren (1777–1835) during his surgical lessons of December 5, 1831 at the hospital Hotel Dieu in Paris (Dupuytren, 1831, 1832). In these lessons, he described a « permanent retraction of the fingers as a result of a crispation in the palmar aponeurosis » (« *retraction permanente des doigts par la suite de la crispation de l'aponevrose palmaire* ») and that « the starting point of the disease was in the exaggerated tension of the palmar aponeurosis, and that this tension itself was due to a contusion of the aponeurosis as a result of the too strong or too long action of a hard body in the palm of the hand » (« *le point de départ de la maladie était dans la tension exagérée de l'aponévrose palmaire, et que cette tension elle-même était due à une contusion de l'aponévrose par suite de l'action trop forte ou trop long-temps prolongée d'un corps dur dans la paume de la main* »). To leave "no doubt in" his students' "mind on the seat of the disease", and "judge… the exactitude of all that" he has advanced, he brought an anatomical specimen showing fingers retraction at a marked degree.[a] He pulled and sectioned different muscles and tendons, but only by sectioning the palmar aponeurosis "the flexion disappears and… the fingers almost return to their normal position" ("*la flexion disparaît et… les doigts reviennent presqu'à leur position normale*"). This was in clear opposition with the prevailing view of those times when it was thought that this condition was secondary to a contraction of the tendons. Napoleon Bonaparte's physician, Alexis Boyer (1757–1833), uses the expression "*crispatura tendinum*" (crispation of the tendons) to describe the hardening and stiffening of the flexor tendons that would result in the contracture (Boyer, 1831; Vrebos, 2009).

---

[a] He wrote: « The chance has fortunately favoured us and we present you the arm, forearm and hand of an individual who has been affected, to a marked degree of finger retraction »; in French, in the original text: "*Le hasard nous a heureusement favorisé et nous vous présentons le bras, l'avant-bras et la main d'un individu qui a été affecté, à un degré marqué, de rétraction des doigts* »

Dupuytren's observation were published in English in *The Lancet* with the title "*Permanent retraction of the fingers, produced by affection of the palmar fascia*" (Dupuytren, 1834). Here, Dupuytren examined the cases of a wine–merchant and coachman and describes how:

> [...]they extend the fingers of the injured hand with less facility than usual; the ring-finger soon begins to contract; the deformity first attacks the extreme phalanx, and the others follow its movement: as the disease advances, the finger becomes more contracted, and the flexion of the two neighboring fingers begins to be remarked (in English in the text)

More details on these clinical cases are furnished in another "Oral Lessons of Clinical Surgery given at the Hôtel-Dieu de Paris" (*Leçons Orales de Clinique Chirurgicale faites à l'Hôtel-Dieu de Paris*), in which he (wrongly) affirms that[b]:

> Most of the individuals affected by this disease have been forced to make efforts with the palms of their hands and handle hard bodies, like hammers, oars and plows [...] We see it in masons who seize stones with the tips of the fingers, among cultivators, etc. We can therefore already see that the disease manifests itself preferably in those who are forced to take use the palm of the hand as a fulcrum.
>
> ***(Dupuytren, 1831, p. 473)***

To support his theory, he describes three clinical cases in which the disease[c]:

> Occurs without having been preceded by any rheumatic or gouty affection, of any inflammation of the tendinous slides, of the articular synovial capsules, without sprain, without distension of the ligaments, without ankylosis, without fracture or external violence of any kind
>
> ***(Dupuytren, 1831, p. 475).***

[b] In French, in the original text: « *La plupart des individus que cette maladie affecte, ont ete obliges de faire des efforts avec la paume de main et de manier des corps durs, tel que le marteau, la rame et la charrue... On voit donc déjà que la maladie se manifeste de préférence chez ceux qui sont obligés de prendre un point d'appui dans la paume de la main. On la rencontre chez les maçons qui saisissent des pierres avec l'extrémité des doigts, chez les cultivateurs, etc. On voit donc déjà que la maladie se manifeste de préférence chez ceux qui sont obligés de prendre un point d'appiii dans la paume de la main* » (Dupuytren, 1831, p. 473).

[c] In French, in the original text: "*Survient sans avoir été précédée d'aucune affection rhumatismale ou goutteuse, d'aucune inflammation des coulisses tendineuses, des capsules synoviales articulaires, sans entorse, sans distension des ligaments, sans ankylose, sans fracture ou violence extérieure quelconque* ».

More details are furnished by Doctors Alexandre Paillard (1803–1835) and Edmon Marx (1797?–1865) in their Report on the Surgical Clinic at the Hôtel-Dieu (Alexandre, 2005; Dupuytren, 1831).[d]

Today, I am only going to talk to you about a single patient and a single illness. The patient I wish to present to you is called Demarteau (Jean Joseph), about 40 years of age and he is a coachman; he actually lies at the [scil. bed] number 63 of the Salle Ste.-Marthe [...] Our patient is a coachman and is therefore obliged to hold in his hand, nearly all the time, a whip with a large and hard handle, and to use it constantly to speed up two bad horses, and the palm of his hand and the palmar surface of his fingers, are thus exposed to a pressure and to a sort of perpetual contusion. I observed this retraction in a wine merchant who is obliged to often taste wines, and during the day has to broach a great number of pieces [scil. casks] with a puncheon; the handle of this punch, large and hard, pressing hard into his palm. I have seen this [scil. contracture] in an office worker who daily sealed many letters with particular care, with wax and a seal whose rounded handle pressed very strongly on the palm of his hand. Finally, I have also seen it in masons and other people who have to lift heavy weights with their fingertips. In other cases, I have not been able to discover any cause for this disease.

This idea was soon challenged by Jean-Gaspard Blaise Goyrand (1803–1866) and Caesar Henry Hawkins (1798–1884) (Hughes, Mechrefe, Littler, & Akelman, 2003). The presence of nodules in the aponeurosis affected by Dupuytren disease was early recognized and the role of these structures in the origin of the contractures was suggested (Skoog, 1948). This role became widely accepted after the description of the presence of myofibroblasts in nodules (see below) (Gabbiani & Majno, 1972; Hinz et al., 2012) (Fig. 3.2).

Despite the eponym, it was one of John Hunter's pupil Henry Cline senior (1750–1877) that first dissected hands displaying features of palmar aponeurosis fibromatosis in 1777 (Dupuytren's birth year) (Osterman, Murray, & Pianta, 2012). In his notes (Fig. 3.3) we read:

[d] In French, in the original text: "*Ainsi, notre malade est cocher de fiacre, obligé, par conséquent, de tenir presque constamment à la main, un fouet a manche gros et dur, de s'en servir sans cesse pour hâter la lenteur de deux mauvais chevaux, et la paume de sa main et la face palmaire de ses doigts, sont ainsi exposées à une pression, et en quelque sorte a une contusion perpétuelle. J'ai observé cette rétraction chez un marchand de vin qui, obligé par état de déguster souvent des vins, est dans l'habitude de donner, dans le cours de la journée, des coups de poinçon a un grand nombre de pièces; le manche de ce poinçon, gros et dur, contondant fortement la paume de la main. Je l'ai rencontrée aussi chez un homme de cabinet, qui cachetait journellement beaucoup de lettres en y mettant un soin tout particulier, avec de la cire et un cachet dont le manche arrondi pressait très fortement la paume de sa main. Enfin, je l'ai remarquée chez des maçons et autres personnes obligées de soulever, avec la pointe des doigts, de pesants fardeaux. Dans d'autres cas, je n'ai pu découvrir aucune cause à cette maladie*».

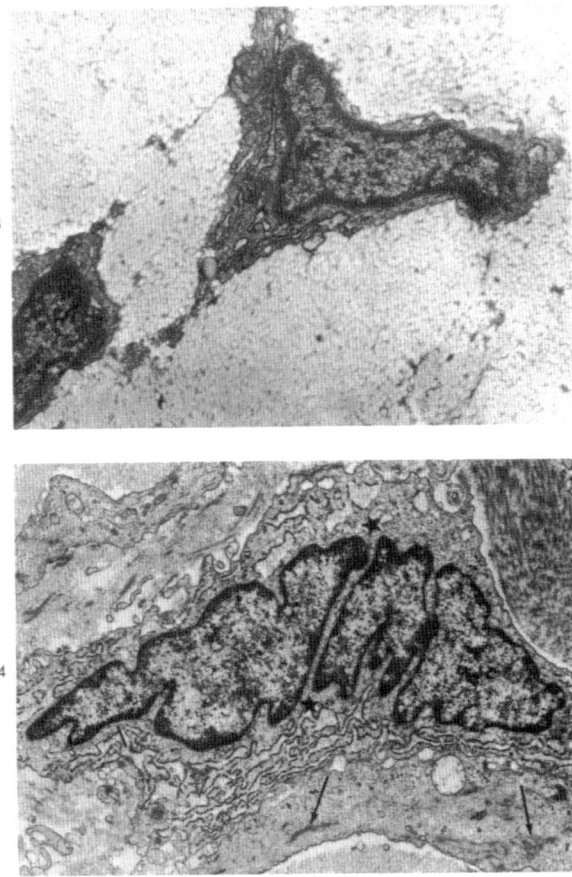

Fig. 3.2 *Upper figure,* 3. Control fibroblast from the palmar aponeurosis of a patient with Dupuytren disease, but in a zone remote from the nodules. Note the relatively smooth outline of the nucleus (× 12,000). *Lower figure,* 4. Fibroblast from a Dupuytren nodule. Note the indented nucleus and the deep folds such as that between the two stars. The cytoplasm contains bundles with dense bodies *(arrows),* (× 10.500). *(Fig. 3 and 4 from Gabbiani, G., & Majno, G. (1972). Dupuytren's contracture: fibroblast contraction? An ultrastructural study. The American Journal of Pathology, 66(1), 131–146. Copyright © 1972 Elsevier Masson SAS. All rights reserved.)*

The contractions of the fingers which so frequently happens in laborious people, arises from a thickening and shortening of the fascia in the palm of the hand, without any alteration in the muscles and tendons. This has been seen in dissecting two subjects, in one all the fingers were contracted, but upon cutting through the fascia they were immediately extended. In the other the ring finger only was contracted, which was found to arise from a thickening and shortening of that portion of ligament, which is inserted into that finger. It appeared from the last case that the fascia is not blended with the skin of the fingers, but inserted into the phalanges distinctly, by the side of it.

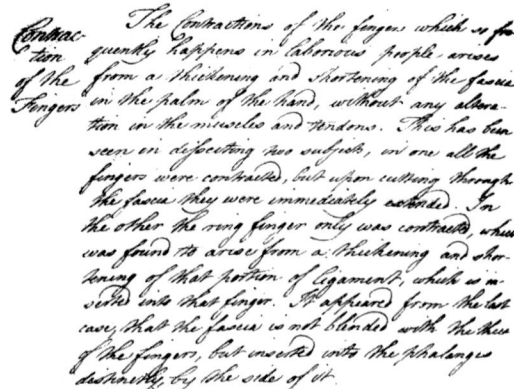

Fig. 3.3 Cline's notes on the "Contraction of the fingers" from the Papers of Henry Cline, 1777–[1824]. *"The contractions of the fingers which so frequently happens in laborious people, arises from a thickening and shortening of the fascia in the palm of the hand, without any alteration in the muscles and tendons. This has been seen in dissecting two subjects, in one all the fingers were contracted, but upon cutting through the fascia they were immediately extended. In the other the ring finger only was contracted, which was found to arise from a thickening and shortening of that portion of ligament, which is inserted into that finger. It appeared from the last case that the fascia is not blended with the skin of the fingers, but inserted into the phalanges distinctly, by the side of it".* (Reproduced by kind permission of the Libraries & Collections at King's College London.)

Sir Astley Paston Cooper (1768–1841), first apprentice, then collaborator and partner of Henry Cline senior, well aware that the disease was secondary to fibrosis of the palmar fascia suggested palmar fasciotomy as a treatment to this type of contraction. In 1822, he writes "… when one aponeurosis (scil. of the palm) is the cause of the contraction, and the contracted band is narrow it may with advantage be divided by a pointed bistoury, introduced through a very small wound in the integument. The finger is then extended, and a splint is applied to preserve it in a straight position" (Cooper, 1851; Hutchison & Rayan, 2011).

Nevertheless, "Dupuytren disease" has a much older history, which dates well before the 18th century. Its first description was done by Felix Platter (or Plater; Felix Platerus, 1536–1614), anatomist, physician (*archiater*) and professor in Basel, Switzerland (but educated physician in Montpellier, France). In his 1614 book "*Observationum in Hominis Affectibus*" (observations in human diseases) he describes a "*contraction digitorum in manibus*" (fingers contraction in the hands), that results in fingers assuming a "starry" conformation (*digiti astrici*) in a famous stonemason (*insignis artifex lapicidia*) (Platter, 1614). See Fig. 3.4.

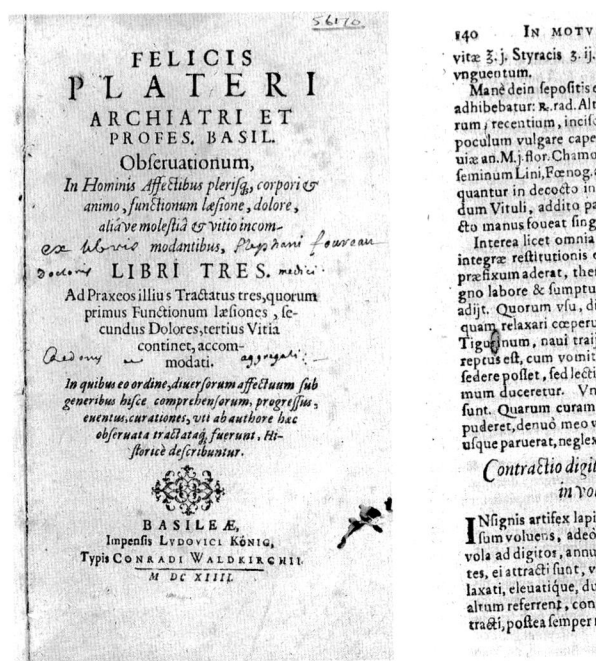

Fig. 3.4 Felix Platter, *"Observationum in Hominis Affectibus"* (Platter, 1614). *Left*: title page. *Right*: detail of page 142, with the description of the contraction of the fingers (*contraction digitorum*), that occurred in a well-known master stonecutter (*insignis artifex lapicidia quidam*) while rolling an enormous stone (*saxum immensum volvens*). *(Credit: Wellcome Collection. Attribution 4.0 International (CC BY 4.0).)*

In his book he describes (our translation from Latin):

"Contraction of the fingers of the left hand. While a well-known master stone-cutter was rolling an enormous stone, the tendons of the palm finishing in the ring- and little-fingers were displaced as if they were freed from their location and protruded at the surface showing two strands stretched under the skin; the two fingers contracted and remained always in the same position."[e] Due to misinterpretation of the original Latin text, it was long thought that Platter misinterpreted the cause of Dupuytren disease as a dislocation and shortening of the flexor tendons. Indeed, Platter proved that the subcutaneous ligamentous extensions of the palmar aponeurosis, *"two strands stretched under the skin"* (duas chordas sub cute tensas, i.e.) were responsible for the contracture.

(Belusa, Selzer, & Partecke, 1995)

[e] In latin, in the original text *"Contractio digitorum sinistrae manus in volam illius. Insignis artifex lapicidia quidam, saxum immensum volvens, adeo tendines in sinistrae manus vola ad digitos, annularem et minimum desinentes, ei attracti sunt, ut illi a vinculi quibus retinetur laxati, elevatique, duas chordas sub cute tensas in altum referrent, contractique duo hi digiti et attracti, postea semper manserint"* (Platter, 1614, p. 142).

Various references suggest that Dupuytren disease was already known before Platter, albeit its anatomical substrate was unknown. Plausible references to the disease exist in Middle Age and 16th century literature from Norther Europe; this accounts for the nickname of Viking disease or Celtic hand (Elliot, 1988). The belief of a Norse origin has recently been disproved by genetic analysis (Ng, Lawson, Winney, & Furniss, 2020). In the 12th and 13th century sagas of the earls of Orkney and of the bishops of Iceland, that relate events that occurred in the 9th or 10th century, are recorded hand contractures compatible with Dupuytren disease that were miraculously cured.

> *"There was a man called Sigurð, from north in Shetland. He had a crippled hand such that all the fingers lay in the palm...", "There was a man called Þorhallr, poor, disabled and old. He had a hand in which the fingers were clenched into the palm, and it had been like this for nearly 60 years; the tendons were knotted so that he could not grasp anything...", "This woman had a hand unusable for work, since three fingers lay clenched into the palm...", "In Papisfjörðr in the East-fjords there was a poor woman whose fingers were disabled: they were clenched into the palm of her right hand."*
>
> **(Elliot, 1988; Flatt, 2001; McFarlane, 2002)**

A Scottish 16th century tale, tells that the bagpipers for the chieftains of Clan Macleod on the Isle Island of Skye, the MacCrimmons, lose their prestigious status when they started to develop a bending of their fingers that made them unable to play the bagpipe (a condition hence called the "Curse of the MacCrimmons") (Elliot, 1988).

It has been hypothesized that the Christian "Hand of Benediction" (*Benedictio Latina*), may have started out of a respect for a pope suffering from Dupuytren disease, probably Sixtus II (?—258 CE), see Fig. 3.5 (Zdilla, 2017). An alternative diagnosis to the peculiar hand gesture would be ulnar nerve palsy (Futterman, 2015; Osterman et al., 2012).

The peculiar posture of the hand is also found in representations of the "Hand of God" (*manus dei*), also called the right (hand) of God/of the Lord (*dextera dei/dextera domini*); see Fig. 3.6.

Some Authors rather believe that the "Hand of Benediction" and well as the "Hand of God", so similar to the Hand of the cult of Sabazios, should be rather interpreted as symbols of victory of the Christian religion over paganism (Barasch, 2000).

Nevertheless, this peculiar gesture existed well before Christianity. The "hand of Sabazios", a votive hand of the cult of Sabazios, the sky-father god of Phrygians and Thracians associated with Jupiter/Zeus or Dionysus, shows features compatible with Dupuytren disease (i.e., flexion of the 4th and 5th digits at the proximal interphalangeal joints, and a cord-like structure visible

Fig. 3.5 Fra Angelico (original name Guido di Pietro, also called Fra Giovanni da Fiesole and Beato Angelico; c.1400–1455). "Sixtus Entrusts the Church Treasures to Lawrence". Detail. Fresco in the Cappella Niccolina of the Palazzi Pontifici in Vatican (1447–49). The fresco depicts the pope Sixtus II (martyred in 258) in the act of blessing Lawrence, and entrusting him the treasures of the church. The pope's right hand displays the typical features of Dupuytren disease, i.e., flexion contractures of metacarpophalangeal and proximal interphalangeal joints associated with hyperextension of distal interphalangeal joint of the 5th finger. The 4th finger only displays mild flexion contracture of the metacarpophalangeal joint. *(Photograph @ Governorate of Vatican City State. Directions of Museums. All Rights Reserved.)*

Fig. 3.6 Mestre de Taüll (Master of Taüll, or Tahul); the hand of God from Sant Climent de Taüll, circa 1123. Museu Nacional d'Art de Catalunya, purchased by the Junta de Museus in the 1919–1923 campaign. The Mestre de Taüll is considered one the most important painters of Romanesque art in Europe, and the greatest fresco painter of the XII century in Catalonia. The fresco shows the hand of God displaying the typical features of Dupuytren disease. *(© Museu Nacional d'Art de Catalunya, Barcelona, 2021.)*

in the palm of the hand), as well as knuckle pads (structures with unknown significance, on the knuckles); see Fig. 3.7 (Zdilla, 2017).

Greek and Roman medical literature seem not to contain records of conditions resembling Dupuytren disease, apart from the *contractures* found in Antyllus' work. Antyllus (or Antyllos), a Greek surgeon who likely lived and worked in Rome between the late third and the early fourth century (i.e., after Galen and before Oribasius, 325–403 CE), although many facts regarding his life are indeed in dispute (see, Grant, 1960). Antyllus work got lost, but Oribasius (physician and friend of the emperor Julian, "The Apostate") refers to Antyllus' work on several occasions, in his 70 (or 72) books anthology (Συναγωγαὶ Ἰατρικαί, *Collectiones Medicae* or Ἑβδομηκοντάβιβλος, *Hebdomecontabiblos*). In particular, he refers to Antyllus' description of contractures (*ankylos*), including hand contractures, that develop as a result of an underlying disease of the tendon (to note: in Antyllus' times the word *nerve* was used to refer to tendons and ligaments); this could indeed be a possible description of Dupuytren disease.(Papadakis, Manios, & Trompoukis,

Fig. 3.7 Hand of Sabazios (*ex voto*). Paris, Musée du Louvre, Inv. No. Br. 836. 3rd century. Sabazios is wearing a Phrygian cap. The scene represented on the wrists represent a woman breastfeeding her child in a cave; this perhaps evokes the birth of Sabazios, son of Jupiter and Persephone. Indeed Persephone, according to Greek's mythology, used to spend half of the year (the autumn and winter season) in the underworld (Hades). This votive hand of the cult of Sabazios shows the pathognomonic features of Dupuytren disease, i.e., flexion contracture of the 4th and 5th finger at the metacarpophalangeal and proximal interphalangeal joints, with sparing of the distal interphalangeal joints. Moreover, a tensed, cord-like structure is visible in the palm of the hand, as well as a nodule in the hypothenar region. The structures on the knuckles could be artistic representations of the knuckle palds (Garrod pads), circumscribed thickenings of the skin over the metacarpophalangeal and interphalangeal articulations that can associate with Dupuytren disease (Zdilla, 2017). (*©RMN-Grand Palais (musée du Louvre)/Hervé Lewandowski.*)

2017) Finally, according to paleopathological investigations, Dupuytren disease could have been present already at the times of mummies. One of the 18 mummies stored in a little tomb used as a warehouse (near to the great tomb of the governor Monthemhat, in El-Assasif, Luxor, Egypt), the mummy M-2 (also called mummy Dupuytren, an adult Theban male, from the Roman period, i.e., 30 BCE–395 CE) presented in his left hand the typical flexion contractures of the 4th and 5th fingers (Dinarès Solà, Baxarias, Fontaine, Garcia-Guixé, & Herrerín, 2012) (Fig. 3.8).

The presence of nodules in the aponeurosis affected by Dupuytren disease (Fig. 3.9) was not recognized by Dupuytren, but was reported by Richer and by Adams (Adams, 1878) respectively, in 1887 and 1878, in their studies of the disease. Adams spoke of "nodular indurations."

Langhans (1887) described an accumulation of proliferating cells and Durel (1888) and Doyen (cited by Durel) called them *"veritable fibromes"* (real fibromas). The existence of nodules was confirmed in the literature through the next decades until the mid of the 20th century without any hypothesis about their function. However, in the nineteen forties and fifties the work by respectively Skoog, Luck and Larsen suggested that the nodules were the first change taking place during the development of the disease and contained fibroblastic cells, which were allegedly responsible for the production of forces resulting in tensed cords within the palmar aponeurosis and hence for finger retraction (Larsen, 1966; Luck, 1959; Skoog, 1948). The role of nodules in the pathogenesis of the disease became widely accepted after the description of the presence of stress fibers containing myofibroblasts in these structures (Gabbiani & Majno, 1972; Hinz et al., 2012).

A contracture of the plantar aponeuroses associated with palmar contracture was first recognized by Dupuytren himself. But it was the German surgeon Otto Wilhelm Madelung (1846–1926) who reported the first isolated case in 1875 (Madelung, 1875). A more detailed description was published in 1894 by another German surgeon, George Ledderhose (1855–1925), to whom this condition is now eponymically associated (Ledderhose, 1894) (Fig. 3.10).

Lapeyronie (or La Peyronie) disease (often misspelled as Peyronie disease), is named after François Gigot de Lapeyronie (or de La Peyronie) (1678–1747), the first surgeon to Louis XV. In 1743 de Lapeyronie described three cases of hard tumors, in the form of nodules or ganglions, in the corpus cavernosum which resulted in painful erections and curvature of the penis on the side of the lesion (de Lapeyronie, 1743). Although he was the first to give a scientific description of this disease, Lapeyronie

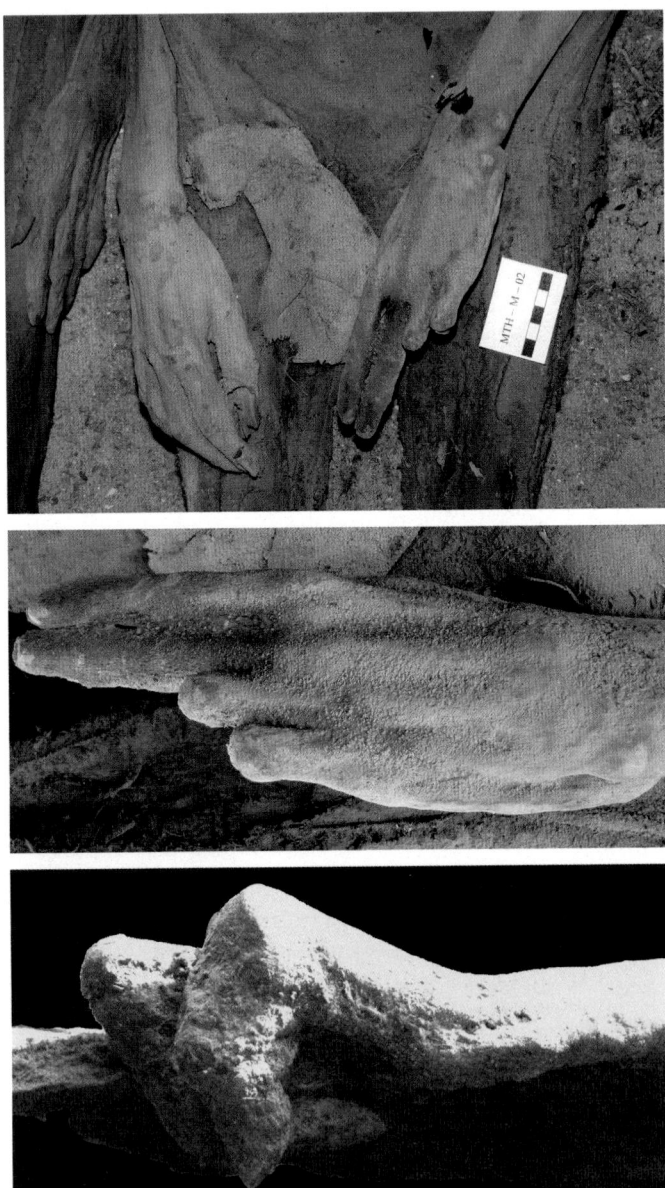

Fig. 3.8 Finger contraction typical of palmar fibromatosis (Dupuytren disease) in the mummy M-2 (also called "mummy Dupuytren"); one of the 18 mummies stored in a little tomb used as a warehouse (near to the great tomb of the governor Monthemhat, in El-Assasif, Luxor, Egypt). The mummy M-2 is an adult Theban male from the Roman period (30 BCE–395 CE); the left hand presents the flexion contractures of the metacarpophalangeal and proximal interphalangeal of the 4th and 5th fingers, with sparing of the distal interphalangeal joint typical of Dupuytren disease (Dinarès Solà et al., 2012). *(Images courtesy of Dr. Rosa Dinares Solà, and Prof Milton Nuñez.)*

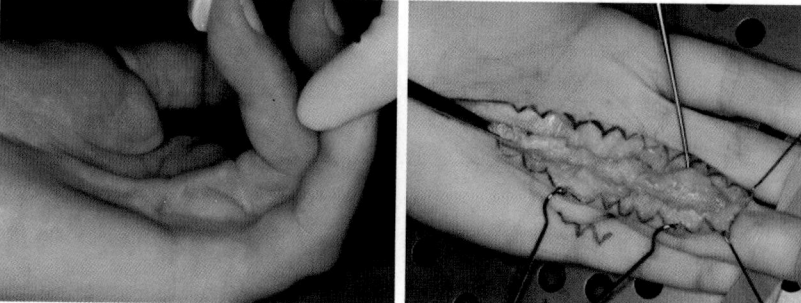

Fig. 3.9 Dupuytren disease. *Left*: View of typical finger retraction in a patient with Dupuytren disease: the aponeurotic cord is clearly visible in the palmar region. *Right*: per-operatory view of the exposed palmar region: now the finger is distended and a large nodule is visible in the palmar aponeurosis, just below the fourth finger. *(Images courtesy of Dr. Michael Papaloizos, Geneva.)*

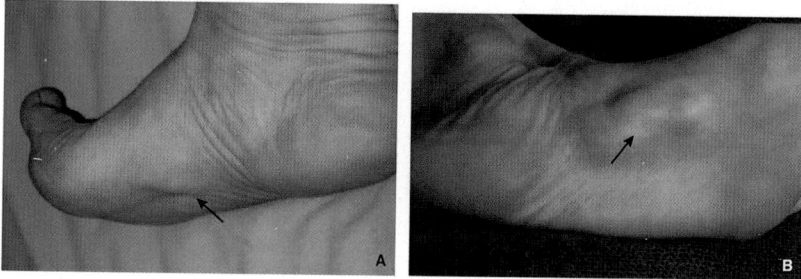

Fig. 3.10 Ledderhose disease. Photographs showing the typical nodules on the plantar area of the feet, localized between the metatarsal heads and the calcaneus. (A) One single nodule *(arrow)*; (B) Multiple, smaller, nodules *(arrow)*. *(Fig. 1 from Damiano, J. (2016). Enfermedad de Ledderhose. EMC Podología, 18(1), 1–5. https://doi.org/10.1016/s1762-827x(15)76063-3. Copyright © 2016 Elsevier Masson SAS. All rights reserved.)*

thought that the indurations typical of the disease that now bears his name developed "*à la suite de quelques accidens vénériens*" ("as a result of some venereal disease"). This penile malformation was mentioned by Guilielmus de Salicetum (circa 1210–1276), Andreas Vesalius (1514–1564), Gabriele Falloppia (1523–1562),[f] Johannes Schenck von Grafenberg (Latin: Ioannes

[f] In his works, Falloppia describes the "ganglions" typical of LaPeryonie disease: "[…] Painless ganglions form, also called "little acorns", which are the cause why, when the penis erects, it swell twisted like a ram's horns, and does not distend. In my opinion this kind of disease is untreatable". In Latin, in the original text "[…] *fiunt ganglia non dolorosa, uel glandulae uocatae, quae postea sunt in causa, ut dum pudendum erigitur ueluti arietinum cornu intortum turgeat, et non distendatur, quod genus morbi mea sententia immedicabile est*" (Falloppia, 1606, p. 169).

Schenckius) (1530–1598),[g] Giulio Cesare Aranzi (or Aranzio, 1538–1589), Claas Pieterzoon Tulp (Nicholaus Tulpius) (1593–1674), and Anton Frederik Ruysch (1638–1731) (Musitelli, Bossi, & Jallous, 2008).

Sir James Paget (1814–1899) in 1879 was the first to link Lapeyronie to Dupuytren disease, nevertheless, he though they both resulted as a consequence of gout [Paget, James. Clinical Lectures and Essays. Gouty affections of Urinary Organs. 2nd, Ed. Page 379. London] This observation was further extended by Fabrizio Gallizia in 1964, who coined the term "*triad collagen*" (collagen triad), to describe the occurrence of Lapeyronie and Dupuytren disease with fibrosis of the auricular cartilage further supporting a common underlying pathogenesis (Gallizia, 1964). We now know that, akin to Dupuytren disease, the characteristic plaques within the tunica albuginea of Lapeyronie disease are composed of fibroblastic cells and altered and disorganized collagen fibers. (Devine Jr. & Horton, 1988; Somers & Dawson, 1997) Albeit often idiopathic, the inflammatory triggers resulting in this fibromatosis, probably originate from vascular trauma or injury.

The knuckle pads, circumscribed thickenings of the skin over the metacarpophalangeal and interphalangeal articulations, were first described by Sir Archibald Edward Garrod (1857–1936), regius professor at Oxford University, better known for his works on the inborn error of metabolism and alkaptonuria (Garrod, 1893, 1904). In his 1904 paper he also foresees an association with Dupuytren disease:

> I have had no opportunity of examining such a pad post mortem, and am therefore unable to give any account of their structure, but their clinical features recall those of thickened bursae, and it is difficult to doubt that they are mainly composed of fibrous tissue… a patient who develops them in early life appears to be liable, as time goes on, to develop Dupuytren's contraction of the palmar fascia.

Indeed, knuckle pads may coexist not only with Dupuytren disease, but also with Ledderhose and LaPeyronie disease, with whom it shares common

---

[g] In his work "Medical observations" (*Observationum medicarum*) (Schenck von Grafenberg, 1596, p. 15), "On a new distortion of the penis" ("*De nova pudendi distorsione*"), he writes (our translation from the original Latin text): "Concerning this disease, I can't find anyone who studied it so to know the real cause of such an effect; nevertheless, they call it nodule of the penis; for when the penis is palpated in its unerected and loose state, one can't feel nothing strange, a part from a little bulge, like a bean or a little acorn, that can indeed be appreciated". In latin, in the original text: "*De hoc morbi genere neminem reperio, qui ita pertractarit, ut veram effectus causam novisse videatur; appellant tamen nodum penis; nam etsi demisso, et flaccido genitali nihil percipiatur incommodi, particulam tamen pertractanti parvulus quidam tumor fabae, vel glandulae speciem referens percipitur.*"

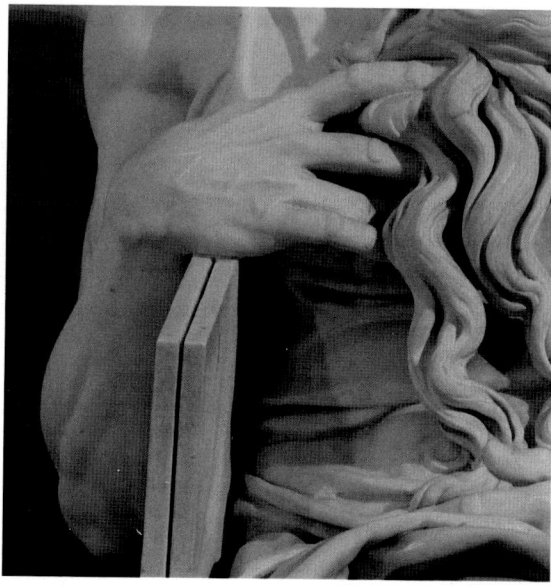

Fig. 3.11 Michelangelo's (Michelangelo Buonarroti 1475–1564), Moses (Italian, Mosé; detail). San Pietro in Vincoli, Rome. The marble statue of Moses is part of the tomb of Pope Julius II, who was commissioned in 1505 by Pope Julius II himself. Moses' hand shows the characteristic skin thickenings over the metacarpophalangeal and interphalangeal articulations, known as knuckle (or Garrod) pads. *(With the permission of the Ministero per i beni e le attività culturali e per il turismo, Soprintendenza speciale Archeologia, Belle Arti e Paesaggio di Roma. Photograph courtesy of Prof. Giovanni Gallo.)*

histological features (Kodama, Gentry, & Fitzpatrick, 1993). Although there is no medical description of knuckle pads before Garrod, visible knuckle pads can be easily identified in Michelangelo's (Michelangelo Buonarroti 1475–1564) statues of David, Moses (see Fig. 3.11) and Giuliano de Medici as well as in God's hand in the creation of the world in the Cappella Sistina (see Fig. 3.12) (Allison Jr. & Allison Sr., 1966).

In all cases the main histological feature of fibromatoses is accumulation of myofibroblasts that in certain locations can result in tissue retraction with important functional implications (Gabbiani & Majno, 1972).

3.3 Scleroderma

Scleroderma, according to a classic definition, is a "generalized disorder of arterioles, microvessels, and diffuse connective tissue, characterized by general scarring (fibrosis) and vascular obliteration of skin, gastrointestinal tract,

Fig. 3.12 Michelangelo, The Creation of Adam (Italian: "Creazione di Adamo"), detail. This fresco painting is a part of the Sistine Chapel's ceiling (painted c. 1508–1512). It illustrates God giving life to Adam by touching with the index finger of his outstretched right arm the index finger of Adam's outstretched left arm. This scene is inspired by the book of Genesis (I.26 and II.7). Interestingly, both God's hand (here only the right hand is shown), but not Adam's show knuckle pad over the interphalangeal articulation of the thumb. *(Photograph @ Governorate of Vatican City State. Directions of Museums. All Rights Reserved.)*

lungs, heart and kidneys, in which hardened skin is the principal clinical trait, and the state of involvement of organs is the keystone in prognosis" (LeRoy et al., 1988). More recently, scleroderma has been considered a complex disease characterized by three principal aspects: vascular alterations, fibrosis and autoantibodies directed against several cellular antigens (Gabrielli, Avvedimento, & Krieg, 2009). Normally, vascular alterations represent the early phase of disease, which is allegedly triggered by an autoimmune process. These alterations of inflammatory nature principally affect small vessels, in particular arterioles (virtually of any organ, but above all of skin and connective tissue) and consist of a loss of endothelial integrity and consequent vascular obliteration. Fibrosis subsequently develops; this coincides with the appearing of the main clinical symptoms. Fibrosis replaces the vascular inflammatory phase, destroying the architecture of affected tissues. This process principally affects the dermis, thus conferring the name to the disease. The skin, because of fibrosis, loses microvascular structures and assumes the characteristics of stiffness and tension of scars.

Two forms of scleroderma are nowadays recognized, one local, known as "limited cutaneous scleroderma", and one systemic, the "diffuse cutaneous scleroderma" (Gabrielli et al., 2009). In the first, fibrosis is limited to hands, arms, and face. Usually, it is preceded by "Raynaud's phenomenon" and can be accompanied by pulmonary hypertension. The second is a disorder rapidly worsening, which affects large areas of the skin and compromises one or more internal organs. The involvement of internal organs is a discriminant factor in prognosis. Kidney, esophagus, heart, and lungs are the most frequent targets. Although the most common visceral complication is severe and debilitating esophageal dysfunction, pulmonary lesions represent the most common cause of death. Reliable data about the epidemiology of disease are still lacking. It is accepted that the incidence of disease varies from 50 to 3.000 cases each millions of individuals, that women are more affected than men are, with a ratio varying from 3:1 to 14:1. The high frequency of autoimmune diseases in the relatives of scleroderma patients can probably be explained by genetic factors. See Fig. 3.13.

Scleroderma represents the most ancient fibrotic disease since his description goes back to the classic times. Cases compatible with a posteriori diagnosis of scleroderma can be found in the classical Greek medical literature such as the work of Hippocrates (660–370 BCE) and Galen (129–216 CE) (Barnett, 1974, p. 4; Pasero & Marson, 2004; Rodnan & Benedek, 1962). In the Aphorisms (V, 71), Hippocrates says "In those persons in whom the skin is stretched, and parched and hard, the disease terminates without sweats; but in those in whom the skin is loose and rare, it terminates with sweats" (Stevenson, n.d.). Whereas Galen, in his book "Hygiene", also known as "On the Preservation of Health" (*De sanitate tuenda*), describes a patient with "constriction of the pores" resulting from obstruction or thickening that occurs together with "white pallor and hardness and thickness of the skin, and during exercise by their difficulty in warming up" (Green, 1951). Nevertheless, there are number of objections to accepting these as description of scleroderma (Rodnan & Benedek, 1962).

Closer to our times, the first clear description of scleroderma is considered the one published in Italian by the Neapolitan physician Carlo Curzio (1692-XVIII cent.) in 1753, translated in English and French within the two following years and widely diffused in Europe (Curzio, 1753). The monograph by Curzio is remarkable because it represents a significant example of the so-called "rational medicine" developed during the second half of XVII century.

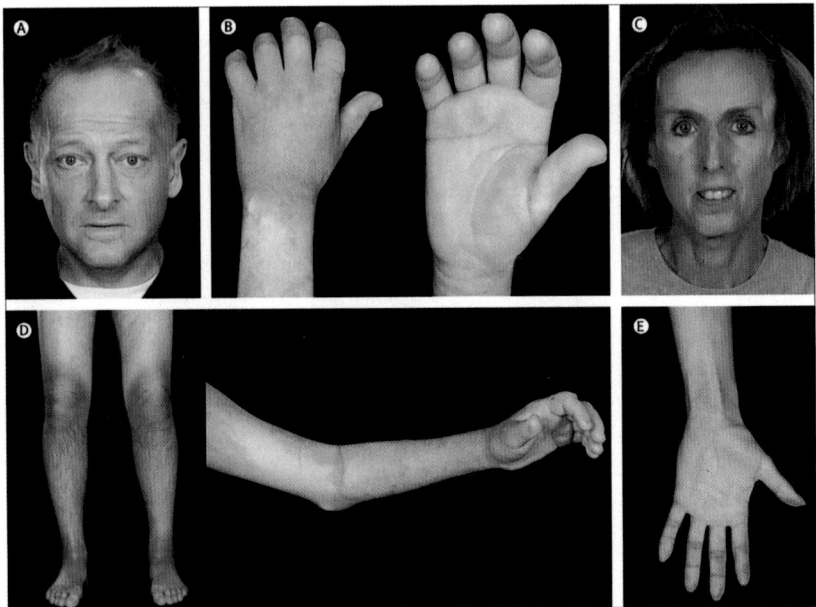

Fig. 3.13 (A) Severe skin involvement in diffuse cutaneous systemic sclerosis has an effect on facial appearance. (B) Hand function is affected in these patients and is often associated with severe digital ulcers and ulceration over areas of pressure or trauma. (C) Atrophic changes of late-stage diffuse skin involvement with prominent hair regrowth. (D) Typical blanching of indurated thickened, hairless skin over the length of the limbs is shown in a patient with early diffuse systemic sclerosis and anti-RNA polymerase antibody positivity. (E) Atrophic changes of the hands in late-stage diffuse skin involvement. *(Fig. 1 from Denton, C. P., & Khanna, D. (2017). Systemic sclerosis. Lancet, 390, 1685–1699. Copyright © 2017 Elsevier Masson SAS. All rights reserved.)*

On June 22nd 1752 Patrizia Galieri, 17 years old, was admitted to the Hospital of Incurables of Naples for an intriguing disease which had made her skin "such as a dry and hard tree bark" (Curzio, 1753, p. III). This news intrigued many physicians and scientists and the Royal Paris Academy asked for a detailed report, which eventually was at the origin of Curzio's book. When Curzio visited the patient for the first time, he immediately noticed an "extreme straightening and hardening of all the skin" (Curzio, 1753, p. VI). This feature was more pronounced on the neck, from which the disease started, and on the forehead, eyelids, lips, tongue and abdomen, at the point that the patient could not move any of these parts, even if, as observed by Curzio, the underlying muscles appeared healthy. The skin, moreover, was barely lukewarm, very sensitive, and without perspiration.

After having presented the case, Curzio envisaged the possibility of an eventual "rational" therapy, based on anatomical and physiological features of the skin. He took into consideration the hypothesis according to which skin's elasticity and softness was due to local gland products and to the processes of transpiration and inspiration, which determined the correct humidification of adjacent tissue layers. The disease, which he defined as a "tonic straightening of the whole skin", could originate from a disorder of the production of "oily" and "aqueous" substances, which guaranteed skin structure and function. This disorder, in turn, could derive from[h]:

> *[...] a tonic contracture of nervous layer, or leathery part of skin, which determined a contraction of excretory ducts of subcutaneous glands and of exhalant vessels [...] by consequence all the body necessarily became rigid, arid, strained and hard.*
> **(Curzio, 1753, p. XL)**

Based on this anatomo-physiological reasoning, Curzio prescribed baths and vapors to re-establish skin's transpiration. He also decided to administer to the patient daily doses of mercury, convinced that the "atoms" of this substance could, in a mechanical way, reopen vascular and glandular conducts inside skin. The patient recovered, a little less than a year after his admission at the hospital.

These explanations and therapeutic strategy may appear speculative, however when they are considered in their historical context they represent a well-advanced medical approach. Curzio, in fact, tried to design a cure based on the up-to-date anatomical and physiological knowledge of his epoch. This knowledge belonged to conception called "iatromechanic", according to which the human body was an engine, more precisely a hydraulic system in which different fluids with different properties flowed, thus regulating health and disease. Glands, in particular, seen as microscopic filters, were thought to have a primary role in formation, transformation and release of bodily fluids. Curzio, as shown by his reasoning about scleroderma, completely adhered to this approach.

In spite of the great interest aroused by the case described by Curzio, his work was not followed by further publications and scleroderma remained not recognized or studied. We have to wait until 1830 to see a new description of the disease and the coining of the term by Giovanbattista Fantonetti

---

[h] In Italian, in the original text: "[...] una tonica contrattura della tela nervea, o parte coriacea della cute, per la quale contraendosi non meno i dotti escretorj delle glandole tutte succutanee, che i vasi esalanti [...] tutto il corpo della cute per necessità dovea irrigidirsi, inaridirsi, stringersi, ed. indurirsi".

(1791–1861), physician in Pavia, who in 1837 published a brief account in the journal "*Effemeridi delle Scienze Mediche*" about a case of "*Skleroderma generale*" (general scleroderma) (Fantonetti, 1837). This term was only in the title, while the article consisted of the clinical presentation about a 30 years old country woman, Antonia Alessandri, hospitalized at the Pavia Medical Clinic on July 7th 1831[i]:

> *[…] all the cutaneous surface was swarthy, the skin was stretched, hard as leather and similar to it […]. Fingers and toes were scarcely movable, as well as arms, hand, and knee articulations […]. It was hard to stick a needle into the skin […] Skin was so hard that it was impossible to cause perspiration, not even with vapour bath.*
> **((Fantonetti, 1837), p. 42)**

Fantonetti, as Curzio, tried to treat the disease with baths and he gave to patient also a hemlock decoction. He thought that skin's "hardening" was caused by an inflammatory process, which determined in turn a "blood blockage" in cutaneous tissue. The disease recovered spontaneously. The scarcity of clinical details could suggest that the disease described by Fantonetti was not really scleroderma, however we are certain that he was the first to introduce the term describing efficiently the skin hardening ("sclero" = hardening, "derma" = skin). Some works have questioned also the case presented by Curzio (Capusan, 1972).

Ten years after the description by Fantonetti, in 1847, few articles were published in France on cases of scleroderma (Barnett, 1974, p. 5). The last of them, which referred to the previous ones, was written by Èlie Gintrac (1791–1877), physician in Bordeaux, who introduced for the first time the term scleroderma in French literature, translated in *sclérodermie* (Gintrac, 1847), favoring its adoption over other alternatives. According to Gintrac the term scleroderma was better than "chorionitis" and "sclérosténose", both introduced in 1847 by Charles Polydore Forget (1800–1861), professor of Medical Clinic at Strasburg. Forget used these terms because he thought that the principal location of the disease was the "chorion", that is, the connective tissue supporting the epithelial membrane (Forget, 1847).

---

[i] In Italian, in the original text: "[…] tutta la superficie cutanea era di colore oscuro, la cute tesa, dura come cuoio, del quale aveva pure l'aspetto […] Le dita delle mani e dei piedi erano appena mobili, e cosi tutte le articolazioni del braccio, della mano e delle ginocchia, nelle quali il tessuto cutaneo risultava meno indurato. Si durava a fatica a conficcare nella cute un ago […] Da che la pelle fu così dura non fu possibile eccitare mai sudore, neanco col. bagno a vapore.

In 1865, the Paris physician Paul Horteloup (1837–1893) published his PhD Thesis "De la sclérodermie", which is considered the most complete treatise of the time. Horteloup reviewed the previous and contemporary literature and produced original interpretations of previous findings. He described 30 new cases, analyzing symptoms, etiology, pathological anatomy, diagnostic, prognostic, and therapy. He compared the skin of his patients to scar tissue as well as to wax, stone, marble or the skin of a frozen cadaver (Horteloup, 1865, p. 57). The involvement of internal organs was not described because no autopsy was performed. Moreover, Horteloup's monograph proposed for the first time a correlation between scleroderma and "Raynaud's phenomenon" described by the French physician Maurice Raynaud (1834–1881), in his degree thesis of 1862 and consisting of vasoconstriction of distal extremities of arms following an intense emotion or cold. Horteloup published the case of a 30 years old farmer previously described by Raynaud, in which he underlined the relationship with scleroderma (Horteloup, 1865, p. 309). In 1899 the British clinician Jonathan Hutchinson (1828–1913) differentiated the Raynaud's phenomenon, characterized by "primitive" spasm of arterioles, from the Raynaud's syndrome, which is secondary to several connective tissue pathologies such as scleroderma and systemic lupus erythematosus (Rodnan & Benedek, 1962, p. 309).

Remarkably, Horteloup advanced that the scleroderma skin lesions were due to a pathological contraction of its "fiber-cells":

> These fiber-cells [...] which formed real little muscles inside the dermis [...] are also part of the structure of arteries and veins, of which they form the muscular layer [...]. If muscular fibers of vascular canals, contracting, diminish a lot their volumes obstructing blood circulation, we can admit that, in the skin, these same fibers could contract to produce a real contraction.

(Horteloup, 1865, pp. 122–123)

Horteloup suggested that this pathological contraction of fiber-cells had a nervous origin, this fact explaining, according to him, why female were more affected than males and why the disease distribution was symmetrical. During the first half of XX century, researchers recognized the systemic nature of scleroderma, which affected internal organs in addition to the skin. The term scleroderma, related to the "dermis", was considered no more correct and the suffix "generalized" was added. An alternative term was introduced, «progressive systemic sclerosis" (Goetz, 1945). Still, the question as to whether one should distinguish a generalized form, from one limited to skin remains open. This latter form (localized scleroderma) is also called "morphea". The British surgeon and dermatologist Sir Erasmus Wilson (1809–1884) has been first credited for using the word morphea in the first half of the XIX century (Rodnan & Benedek, 1962).

In the XIII Chapter (Disease from Special Internal Causes) of his "On diseases of the skin" (Wilson, 1847) he writes: "[…] Morphea, derived from the Greek word μορφή, forma, signifying a visible appearance, and in application to its seat, a visible appearance or spot upon the skin, is, as its name implies, a spot upon or in the skin, of irregular form […]"

He the distinguishes two principal forms of morphea:

> […] white morphea alba, or of dark-brownish or blackish hue, morphea nigra. Moreover, morphoea alba admits of a secondary division, from presenting two varieties, one in which there is induration of the skin, from deposition in its tissue of a lard-like substance, morphea alba lardacea, vel tuberosa; the other being distinguished by atrophy of the skin, and by a greater degree of insensibility, morphea alba atrophica, vel anesthetica.[…]

Although the accurateness of his description, Wilsons considers morphea as a pathognomonic sign of leprosy (or elephantiasis).

> […] elephantiasis having for its synonyms, lepra Arabum, lepra Judaeorum, and lepra medii sevi; bucnomia, or Barbadoes leg, having for its synonym, elephantiasis Arabum; and lepra, for its synonym, lepra Groecorum […] that grand, that elephant disease, the leprosy of the middle ages, which forms so prominent a feature in the history of Europe, and especially of Great Britain, of which examples have not very long vanished from our land […] Although a mere shadow in comparison with the parent disease, it is nevertheless sufficient to occasion considerable annoyance to the sufferer, and to bring him not unfrequently under the inspection of the medical man. Nor, when once pointed out, can the medical man doubt for an instant the nature of the disease which he has before him: there is the insensibility, the deposition, the blanching, the exhaustion of function, and the atrophy of the parent malady, with all their original distinctness, indeed, one complete symptom of the pure elephantiasis, preserved unchanged, as it existed among the Jews, and as it is to be found at this moment on the shores-of Norway, the symptom which was called by the ancients morphoea.[…]

In this period, the concept of "collagen disease" was proposed for the first time in order to define a group of diseases characterized by a dysfunction of the connective tissue extracellular matrix, such as "rheumatic fever", "lupus erythematosus" and "scleroderma", and characterized by a "fibrinoid alteration". This idea represented an important conceptual innovation, characterizing a family of diseases that had escaped until then any classification. Rheumatic fever was discovered to be characterized by fibrinoid changes and then this concept was applied to lupus and scleroderma, two "systemic diseases affecting essentially connective tissue. In turn, connective tissue was conceived as the system exerting the physiological function of favoring the transit and transfer of metabolites and of setting the hydrosaline equilibrium of the body (Klemperer, Pollack, & Baehr, 1984). These diseases, according to the scientists of that time, were characterized by three pathological processes: fibrinoid degeneration, proliferative changes, causing densification of connective tissue, and inflammatory reaction. These features were at the basis of the definition of "collagen diseases", each specific disease being characterized by a different degree and a combination of the three processes and by a different anatomical distribution (Duff, 1948). The concept of "collagen diseases", however, soon revealed its limits (Klemperer, 1950), and was abandoned. Today the term "collagenopathy" is limited to diseases specifically of the collagen molecule.

In 1964 Tichard Winterbauer described a syndrome found in eight patients composed by calcinosis, Raynaud's phenomenon, sclerodactyly, and telangiectasia (Winterbauer, 1964). Some years later, to this picture was added the presence of esophageal lesions, and the syndrome was given the acronym CREST (calcinosis, Raynaud's phenomenon, esophageal dysmotility, sclerodactyly, and telangiectasia). This syndrome was considered a subset of scleroderma. In 1988 two fundamental forms of scleroderma were distinguished: the "Diffuse Cutaneous Systemic Sclerosis" (dSSc) and the "Limited Cutaneous Systemic Sclerosis" (lSSc), this latter including the subset of CREST (LeRoy et al., 1988). This classification is still used today (Gabrielli et al., 2009).

In the 1970's experimental work suggested the hypothesis that the fibroblast was involved in the pathogenesis of systemic sclerosis. The excess of fibrous tissue was explained, at the beginning, as an excess of collagen secretion. It was demonstrated that fibroblasts from patients affected by systemic sclerosis showed in vitro an excess of collagen synthesis as compared to control cells (Leroy, 1972).

Following the discovery of the myofibroblast and of his role in fibrotic diseases, the presence of this cell in scleroderma lesions was investigated and confirmed both in local and systemic scleroderma. Moreover, it was suggested that this cell plays an important role in the onset of collagen accumulation during scleroderma (Sappino, Masouyé, Saurat, & Gabbiani, 1990). Myofibroblast and collagen accumulation correlates well with the typical presence of inflammatory changes in the dermis of scleroderma patients. The possibility that excess of fibrosis could be due to a defect in the process of myofibroblast apoptosis was also suggested (Rajkumar et al., 2005) (Fig. 3.14).

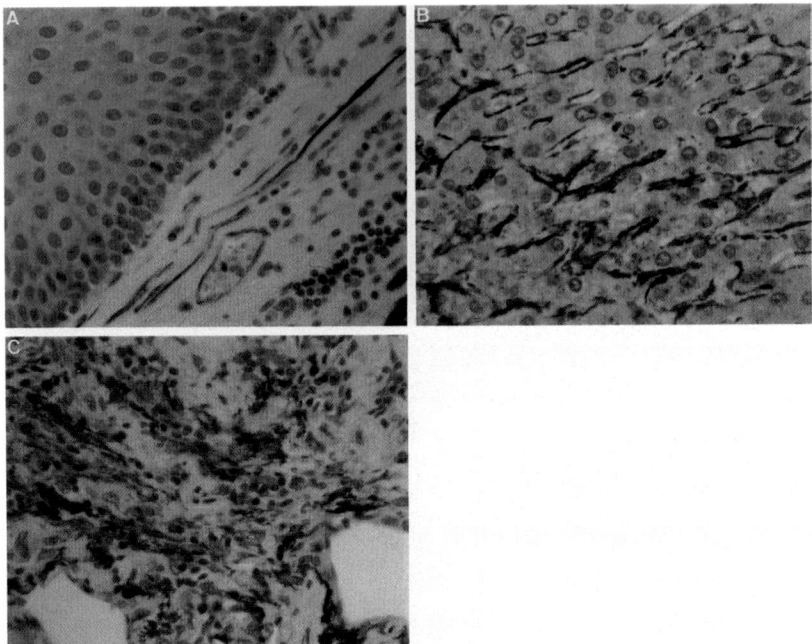

Fig. 3.14 Immunoperoxidase staining with anti-α-SMA antibodies of esophagus (A), liver (B) and lungs (C) tissue sections from progressive sclerosis autopsy specimens. In all cases, stromal cells appear strongly positive. In the liver (B) no fibrosis is present but sinusoidal cells appear strongly reactive. (A) and (B): ×1200; (C): ×600. *(Fig. 3 from Sappino, A. P., Masouyé, I., Saurat, J. H., & Gabbiani, G. (1990). Smooth muscle differentiation in scleroderma fibroblastic cells. The American Journal of Pathology, 137(3), 585–591. Copyright © 1990 Elsevier Masson SAS. All rights reserved.)*

Many efforts are displayed in order to unravel the mechanisms controlling the excess of collagen expression during scleroderma in view of possible therapeutic strategies. Recently, it has been shown that CD38 (a multifunctional enzyme involved in cellular energetics) is upregulated in the skin of scleroderma patients and that targeting CD38 results in reduction of collagen expression in cultured fibroblasts and in experimental lung fibrosis in mice. CD38 may represent a promising target for controlling fibrosis evolution (Kuppe et al., 2021).

3.4 Myocardial fibrosis

Both in the Ancient and in the New Testament, the word sklerokardia (i.e., "hardened heart") is used in a metaphorical, spiritual sense to define one who is destitute of spiritual perception. The term "cardiac sclerosis" would acquire a pathological sense only much later.

According to the humoral theory of Greek classic medicine a disorder in part of the body could be determined by an excess or a defect of a given humor, or of a given "quality" (namely hot, cold, dry, or humid) correlated with a humor. In that theoretical context, we find one of the first mentions of a pathological "*dryness*" of the heart, which evokes a fibrosis of the organ. In his *On the prognosis by the pulse*, Galen related mental states as worry and grief (comparable to nowadays "stress") to heart "dryness", and pulse "hardness". Conversely, idleness and excess of flood could make the heart excessive "moist" and the pulse "soft": "*Starvation, as during long detention in ports, immoderate toil, old age, worry, grief and prolonged loss of sleep, also hectic fevers [...] make the heart dryer and the pulse harder than normal. But a moist dyscrasia is caused by the prolonged occurrence of longstanding edema and watery leukophlegmasia which in turn modify the heart; also by edema of the principal organs. All these make the arteries and the heart moist and the pulse soft. [...] The cause of this can be ingestion of too moist food, idleness, prolonged sleep and bathing after meals*" (Siegel, 1968, p. 336).

A brief history of cardial fibrosis can be found in the doctoral thesis of the French physician and microbiologist Maurice Nicolle (1862–1932), "*Contribution à l'étude des affections du myocarde. Les grandes scléroses cardiaques*" (Contribution to the study of myocardial disorders. Major cardiac sclerosis). Much interestingly, in the incipit he states that "*Of all the visceral scleroses, that of the heart is perhaps the most poorly known*", and attributes this to a lack of anatomical basis (Nicolle, 1890, p. 7). Moreover, that myocardial fibrosis might be determined by several different causes and it is characterized by nonspecific symptoms represented a further obstacle to its full understanding.

Nicolle divides the history of myocardial fibrosis in three phases. The first characterized by the first morphological and macroscopic descriptions based on autopsy, but without any clear understanding of the pathology involved. Nicolle mentions authors such as the Italian physician and anatomist Ippolito Francesco Albertini (1662–1738), the famous Dutch clinician Hermann Boerhaave (1668–1738) and few others who, in his opinion, gave only brief and uncertain descriptions which cannot be considered (Nicolle, 1890, p. 8). According to Nicolle, the first account of myocardial fibrosis was done by the Italian anatomist Giovanni Battista Morgagni (1682–1771). By the way, Albertini was his teacher of anatomy when Morgagni was student at Bologna University, while Boerhaave was his close friend (Zampieri, 2016). In the 45th letter, paragraph 23, of his famous *De sedibus et causis morborum per anatomen indagatis*, he reports the autopsy of a 40-year-old woman, who died in one or two hours without ever experiencing palpitations during her life. The autopsy revealed a hypertrophy of the left ventricle with sclerotic plaques in the wall, pillars, and interventricular septum (interestingly, the aorta was atheromatous and dilated). Morgagni stated that "The defect of the muscle of the heart, degenerating in a tendinous nature, was more evident from the internal to the external side of the left ventricle"[j] (Morgagni, 1761). When discussing his anatomo-clinical observations, Morgagni was always keen in reporting possible previous descriptions made by ancient or modern authors. Given that in this case he did not mention any other source, we might be reasonably sure that this is the first original account of a myocardial fibrosis, even if Morgagni gave only a pure morphological description without any further information or hypothesis.

Nicolle closes this first period mentioning some of the most important protagonist of the French anatomo-clinical school, flourished after French revolution, such as Jean-Nicolas Corvisart (1755–1821),[k] René Laennec (1781–1826),[l] Gabriel Andral (1797–1876), and Jean Cruveilhier (1791–1874).[m] However, in all these cases, the descriptions are poor both from an anatomical and from a clinical point of view. They lack any attempt of etiopathogenic explanation (Nicolle, 1890, p. 8). Finally, Nicolle mentions that the term "myocarditis" was first introduced by the German physician Joseph Friedrich Sobernhein (1803–1946) in 1837 (Sobernheim, 1837); the term "cardiac cirrhosis" was coined by the English physician John Syer Bristowe (1827–1895) in 1856; while the Austrian physician Josef Hamernjk (1810–1887) in 1843 understood that the lesion was mostly found in the left ventricle (Nicolle, 1890, pp. 8–9).

The second phase of this history was characterized by some advancements not in relation to the myocardial fibrosis in itself but related to other conditions indirectly connected to that one. He mentions, in particular, the works of the French physicians Maurice Letulle 1853–1929) and Maurice Debove (1845–1920), and the German physician Karl Weigert (1845–1904). Letulle observed a sclerosis that results from cell proliferation,

---

[j] In Latin, in the original text: "Vitium hoc carnis cordis, in tendineam naturam degenerantis, quo magis ab interiore ventriculi facie ad exteriorem pergebat, eo hebat evidentius"

[k] As well known, Corvisart developed chest percussion as a diagnostic tool for chest diseases. He was personal physician of Napoleon Bonaparte (1769–1821).

[l] Laennec, pupil of Corvisart, is universally known for having invented the stethoscope, an instrument that allowed him to diagnose heart and lung disease in living patients.

[m] Andral and Cruveilhier are important for their contribution to pathological anatomy. Chruveilhier, in particular, published the *Anatomie pathologique du corps humain* (1829–1842), which is one of the first texts of pathology illustrated (there are over 200 copper plates created by the artist Antoine Chazal).

either perivascular or perifascicular, in the hearts of individuals with valve lesions (Letulle, 1879). Letulle and Debove, moreover, mentioned cardiac sclerosis as a consequence of interstitial nephritis (Letulle, 1880). Weigert recognized that sclerosis of the myocardium resulted always in the obliteration of the coronary arteries. While the abrupt suppression of blood flow leads to the production of a necrotic focus in the myocardium, its quantitative reduction causes the disappearance of muscle fibers and the subsequent formation of *"calluses"*. He was one of the first to understand the causal relation between coronary occlusion and myocardial infarction (Hort, 2002).

Finally, the third phases, which finishes with Nicolle's reconstruction at the end of the 20th century, is characterized by specific studies on myocardial sclerosis in itself, in particular from a pathological perspective, with also some insights from the etiological and clinical point of view (Nicolle, 1890, pp. 10–17). With regard to the etiological perspective, it might be worth mentioning the work of the Swiss pathologist Enrst Ziegler (1849–1905), who advanced that the majority of cardiac sclerosis is ischemic in origin, but he also suggests an inflammatory origin.[n] Nicolle stated that his histological descriptions were the most advanced at disposal at that time.

Finally, with regards to etiology and clinics, Nicolle noted that already in 1860 the famous German pathologist Rudolph Virchow (1821–1902) observed that syphilis was probably responsible for the production of certain types of myocarditis. Other authors, in the late 19th century, established connection between smallpox, scarlet, typhoid, and rheumatic fever and sclerotic conditions of the myocardium (Nicolle, 1890, pp. 16–17). At the end, he summarized that that three kinds of cardiac sclerosis have been described: some resulting from known infectious causes, others related to secondary hypertrophy, and the last one inseparable from arteriosclerosis.

The concept of myocardial fibrosis has started to emerge at the end of the 19th century thanks to Authors from Germany, France and United States (Evans & Nuzum, 1929; Mac Callum, 1889). Fibrosis accompanies a large number of myocardial pathologies and has been shown to be instrumental in the development of cardiac dysfunction in the failing heart (Delaunay, Osman, Kaiser, & Diviani, 2019). The myocardial pathologies evolving toward fibrosis are characterized by early cardiomyocyte death,

---

[n] Ziegler was the author of a famous treatise of pathology, the *Lehrbuch der allgemeinen und speciellen pathologischen Anatomie und Pathogenese* (1882) (Ziegler, 1882), which was translated in English ("A Text-Book of Pathological Anatomy and Pathogenesis", 1883–1884) and widely used in Europe.

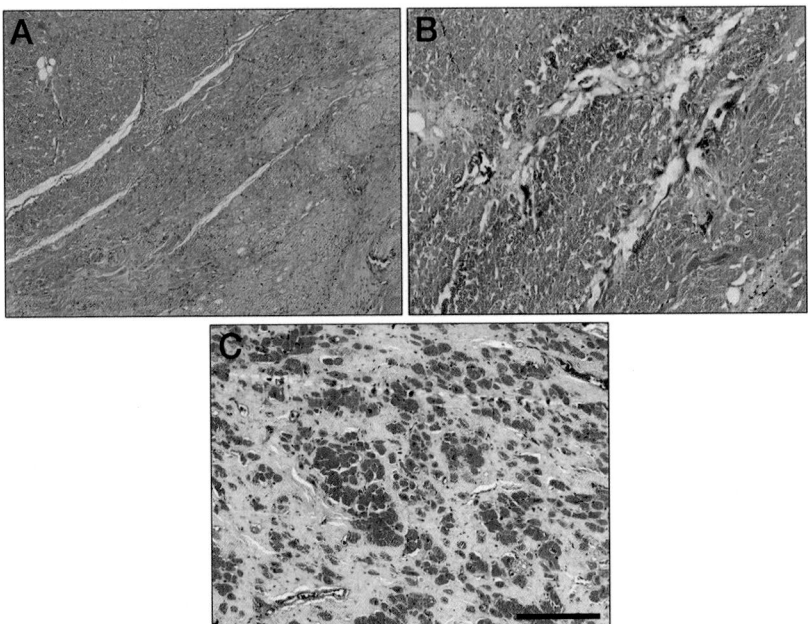

Fig. 3.15 Movat-Pentachrome staining of the adult human ventricle. (A) Healthy adult, (B) ischemic cardiomyopathic (interstitial fibrosis) and (C) dilated cardiomyopathic (diffuse fibrosis) muscle tissue. Muscle tissue is red, proteoglycans and glycoproteins are shown in blue green, and collagens are depicted in *yellow*. Scale bar equals 200 μm. *(Fig. 2 from Hinderer, S., & Schenke-Layland, K. (2019). Cardiac fibrosis – A short review of causes and therapeutic strategies. Advanced Drug Delivery Reviews, 146, 77–82. https://doi. org/10.1016/j.addr.2019.05.011. Content licensed under the Creative Commons Attribution 4.0 International (CC BY 4.0).)*

e.g., after myocardial infarction (also called replacement fibrosis) or cardiac damage without evident cardiomyocyte death, e.g., chronic myocardial insufficiency (also called reactive fibrosis). In both cases the progressive accumulation of extracellular matrix, consisting essentially in heavily cross-linked collagen type I, aggravated by the appearance of perivascular fibrosis, results eventually in cardiac failure (Talman & Ruskoaho, 2016) (Fig. 3.15).

The lesions appearing during replacement fibrosis resemble those evolving during classical skin wound healing with the formation of a scar and impairment of cardiac function due to the loss of myocardial tissue. In addition to these changes, myocardial fibrosis accumulates gradually far from the ischemic lesion, by reactive fibrosis. In this case, as well as in other more chronic cardiac diseases, fibrosis contributes to the establishment of cardiac insufficiency (Gullberg, Kletsas, & Pihlajaniemi, 2016).

Much progress has been made in the last decade in the understanding of the mechanisms leading to the reactive myocardial fibrosis establishment and now an acceptable scheme of the sequence of events conducing to this lesion can be envisaged: the initial change appears to be the focal death of myocardial cells, which in turn elicits a discrete local accumulation of inflammatory cells that is followed by myofibroblast differentiation with synthesis and deposition of extracellular matrix (Weber, Sun, Bhattacharya, Ahokas, & Gerling, 2013). Obviously myocardial death has different causes according to the pathological situation, but the resulting inflammatory sequence is similar in all cases and promotes the same fibrotic consequences. The cardiomyocyte is a fragile cell and it is not surprising that a variety of noxious events produce a necrosis of this cell: due to its relatively low regenerative activity, an inflammatory reaction takes place, which ends in fibrosis accumulation with more or less granulation tissue formation and more or less myofibroblast differentiation, as testified by α-SMA expression, according to the clinical situation (Weber et al., 2013).

Much effort has been spent in order to establish the origin of fibroblastic cells during myocardial fibrosis. Apparently most fibroblastic cells of both reactive and replacement fibrosis derive from resident fibroblast populations (Moore-Morris et al., 2014). This may be at variance with other organ fibroses in which phenomena such as epithelial- or endothelial-mesenchymal transition as well as migration of bone marrow derived circulating cells contribute to myofibroblast accumulation (Tomasek, Gabbiani, Hinz, Chaponnier, & Brown, 2002).

Despite the important advances in the understanding of the mechanisms leading to myocardial fibrosis establishment, no real progress has been made in the finding of efficient therapeutic strategies, however many laboratories are working in order to finding therapeutic tools against myocardial and other fibroses, and it is possible to foresee progress in this direction in the near future. For a review see (Fang, Murphy, & Dart, 2017; Webber, Jackson, Moon, & Captur, 2020).

Compounds with a direct anti-fibrotic effects (i.e., the antagonists of the Connective Tissue Growth Factor, CTGF, and the Galectin-3, Gal-3, inhibitors) have been tested in the animals with interesting results, but their effect on human myocardial fibrosis is still not proven (Webber et al., 2020). In particular, the CTGF is a matricellular protein modulating the signaling pathways responsible for myofibroblast activation and underpinning the pathogenesis of fibrosis (Lipson, Wong, Teng, & Spong, 2012); Gal-3, instead, regulates transdifferentiation of quiescent fibroblast into myofibroblasts

(Henderson et al., 2008). Silencing MicroRNAs (miRNAs), i.e., small non-coding RNA molecules that can control gene expression by silencing the gene, are expressed by cardiac myocytes that regulate signaling pathways in fibroblasts. In vivo, silencing miRNAs can prevent interstitial fibrosis (Crunkhorn, 2009; Webber et al., 2020). Renin-Angiotensin-Aldosterone System Inhibitors drugs (e.g., lisinopril, losartan), as well as mineralocorticoid receptor antagonists (e.g., spironolactone) can target cardiac fibrosis, independently from their antihypertensive effects (Fang et al., 2017; Shearer, Lang, & Struthers, 2013). Likewise, loop diuretics (e.g., torsemide), another common therapy for the treatment of heart failure, can affect collagen deposition (Buggey et al., 2015).

Transforming Growth Factor-β (TGFβ) is a major profibrotic cytokine. TGFβ inhibitors (e.g., pirfenidone and tranilast) can, at least in the animal model, reduce fibrosis and improve functional outcomes (Webber et al., 2020). Downstream of TGFβ, the inhibition of IL-11, a major regulator of fibrosis (Schafer et al., 2017), can exert important effects against the development of myocardial fibrosis, at least in the animal model.

Although blockade of the endothelin receptors (e.g., bosentan) could also potentially prevent cardiac fibrosis, data are discordant and inconclusive.

A possible, indirect anti-fibrotic effect of the inhibitors of the matrix metalloproteases (MMPs) and of the vasodilator hormone relaxin has been postulated; yet, further research is needed.

Although drugs with anti-inflammatory properties, like the statin rosuvastatin (Zhang et al., 2012), the TNFα antagonist etanercept and infliximab (Rolski & Błyszczuk, 2020), and the peroxisome proliferator-activated receptor (PPAR) agonists (Li et al., 2018) could modulate the development of myocardial fibrosis, their clinical benefit is still debatable.

Quite recently, the finding that osteopontin plays an important role in the establishment of reactive myocardial fibrosis suggests that this protein represents a promising therapeutic target (Abdelaziz Mohamed, Gadeau, Hasan, Abdulrahman, & Mraiche, 2019).

3.5 Kidney fibrosis

Renal fibrosis represents the inevitable outcome of chronic renal injury. The term "nephrosclerosis" ("kidney hardening") was coined in 1918 by the German clinicians and pathologists Franz Volhard (1872–1950) and Theodor Fahr (1877–1945) (Meyrier, 2015). The etymology refers to the callous consistence of kidneys cut after removal at autopsy. Hardening suggests tissue fibrosis and can apply to most chronic kidney diseases but Volhard distinguished three variants of "nephrosclerosis," the third one linked to hypertension. The clinician, Franz Volhard, and the pathologist, Theodor Fahr, worked closely together in Mannheim from 1909 until 1915 and introduced a novel classification of renal diseases. In their 1914 monograph entitled «Bright's kidney disease. Clinic, Pathology and Atlas" (*Die Brightsche Nierenkrankheit. Klinik, Pathologie und Atlas*) (Volhard & Fahr, 1914) they differentiated between degenerative (nephroses), inflammatory (nephritides) and arteriosclerotic (scleroses) diseases. To note, they used the eponym "Bright disease" (after the English physician Richard Bright, 1789–16 December 1858) to refer to a chronic kidney disorder characterized by oedema (dropsy), proteinuria (coagulable urine) and uremia (end stage renal disease) (Peitzman, 1989), as reported by Bright in his 1836 article "Cases and Observations Illustrative of Renal Disease Accompanied with Secretion of Albuminous Urine" (Bright, 1836). Indeed Volhard and Fahr's book, is considered "one of the milestones of medical history'" because of its pathogenetic classification, based on both clinical and pathological anatomical findings (Mahon, 1966).

Fibrosis may affect both the glomerulus and the remaining parenchymal tissue, resulting in glomerular and tubule-interstitial sclerosis with abnormal accumulation of extracellular matrix (ECM), thus contributing to glomerular and tubular dysfunction and eventually to renal insufficiency (Sun, Qu, Caruana, & Li, 2016). Myofibroblasts appear as the main responsible for excessive ECM accumulation in the glomerulus, where mesangial cells are allegedly the main source of myofibroblasts while in the interstitium a number of local and possibly circulating cells, such as local fibroblasts, pericytes, epithelial and endothelial cells (through epithelial- and endothelial-mesenchymal transition), as well as fibrocytes have been proposed to contribute to myofibroblast origin (Fierro-Fernández et al., 2020; Sun et al., 2016). The local origin of myofibroblasts in the interstitium has been recently confirmed by single-cell RNA sequencing; this technique has also indicated that naked cuticle homologue 2 represents a possible therapeutic target for interstitial kidney fibrosis (Kuppe et al., 2021). As in other organ fibroses, the most accepted mechanism of myofibroblast differentiation involves TGF-β through the Smad dependent and independent signaling pathways.

Although a number of potential anti-fibrotic treatments that could specifically target kidney fibrosis, no therapeutic tools have been yet developed in order to counteract the development of kidney fibrosis (Fierro-Fernández et al., 2020; Klinkhammer, Goldschmeding, Floege, & Boor, 2017; Ruiz-Ortega, Rayego-Mateos, Lamas, Ortiz, & Rodrigues-Diez, 2020) (Fig. 3.16).

3.6 Pulmonary fibrosis

The concept of idiopathic pulmonary fibrosis (IPF) started to emerge with the systematic clinical and radiological work by the internist Louis Hamman and the pathologist Arnold Rich at the beginning of the thirties of last century (Hamman & Rich, 1944; Noble & Homer, 2005). The discussion of pathogenesis of IPF started with their description of four patients who succumbed to respiratory insufficiency between 1931 and 1943 at the Johns Hopkins Hospital (Noble & Homer, 2005). Interestingly, Ludwig Von Buhl (1816–1880) (Von Buhl, 1872) introduced the term of "muscular cirrhosis of the lung" (*Müskulare Zirrhose des Lunges*) to describe those cases of pulmonary fibrosis characterized by hypertrophy of the muscular layer of respiratory bronchioles, resulting in sclerosis of the pulmonary interstitial tissue with dilatation of the alveoli and terminal bronchioles. The lungs finally acquire a "hobnail" appearance, similar to that of Laennec's cirrhosis. The fibrotic lesions associated with air-containing micro-cysts (i.e., dilated bronchioles) prompted the German pathologist Georg Eduard von Rindfleisch (1836–1908) to use the term "Lung Cystic Cirrhosis", in Latin: *Cirrhosis cystica pulmonum* (Rindfleisch, 1897) (Fig. 3.17).

As most other fibrotic lesions, IPF appears to result from an inflammatory process within the lung alveoli, in this case without a known pathogenesis, evolving in long lasting fibrotic changes determining the poor evolution of the disease (Noble & Homer, 2005). Most probably, the lesion initiates with repeated damage of epithelial cells, alveolar type 2 epithelial cells in particular (Nemeth, Schundner, & Frick, 2020). During the last decade progress about the understanding of IPF pathogenetic mechanisms has been made: a combination of genetic predisposition, ageing and environmental exposure appear to be responsible of the initiation and progressing of the disease (Richeldi, Collard, & Jones, 2017; Yanagihara, Sato, Upagupta, & Kolb, 2019). The participation of autoimmunity processes, characterized by inflammatory infiltrates with mononuclear cells, has been also suggested (Hoyne, Elliott, Mutsaers, & Prêle, 2017). The presence of contractile proteins, α-SMA in particular, in fibroblastic cells (Evans, Kelley, Low, & Adler, 1982; Mitchell et al., 1989; Vyalov, Gabbiani, & Kapanci, 1993) as well as increased contractility of fibrotic tissue (Mitchell et al., 1989) during experimental pulmonary fibrotic changes and human IPF (Kapanci, Desmouliere,

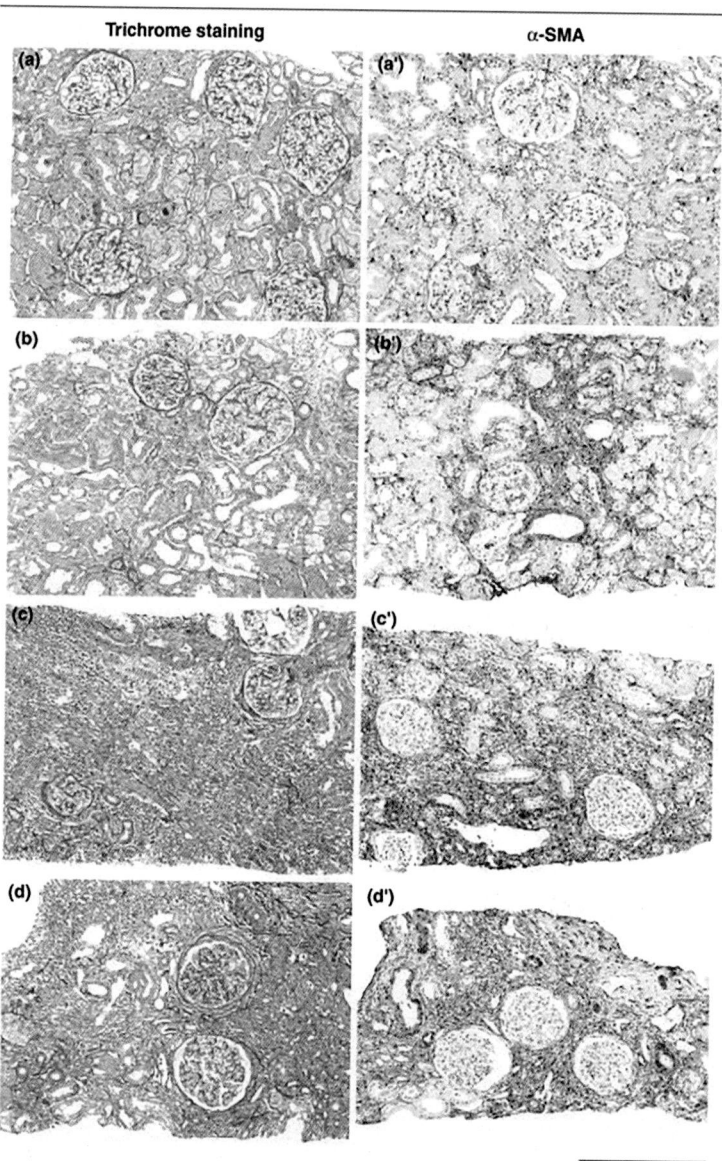

| Trichrome staining | α-SMA |

Fig. 3.16 Interstitial fibrosis: a common pathological feature observed in any type of chronic kidney disease (CKD) irrespective of the etiology. Graded renal biopsies from minor to severe CKD patients. Different methods of staining (A–D, trichrome staining; A′–D′, α-SMA immunostaining) enable the researcher to understand the evolution of organ remodeling during disease progression. Biopsies from patients with low grade fibrosis (A and E) show almost preserved epithelial tubular cell architecture and a low amount of extracellular matrix (ECM), labeled in orange and blue respectively, in (A). In these patients, α-SMA expression is limited to vessels (A′). In patients with severe fibrosis (D and D′), ECM, produced by activated myofibroblasts, accumulates at the place of functional tubules (seen by comparing the orange vs blue staining ratio, see (D), and α-SMA expression, see (D′), is upregulated, revealing activation of resident myofibroblasts. *(Fig. 2 from Prunotto, M., Gabbiani, G., Pomposiello, S., Ghiggeri, G., & Moll, S. (2011). The kidney as a target organ in pharmaceutical research. Drug Discovery Today, 16(5), 244–259. https://doi.org/10.1016/j. drudis.2010.11.011. Copyright Elsevier Masson SAS. All rights reserved. © 2011.)*

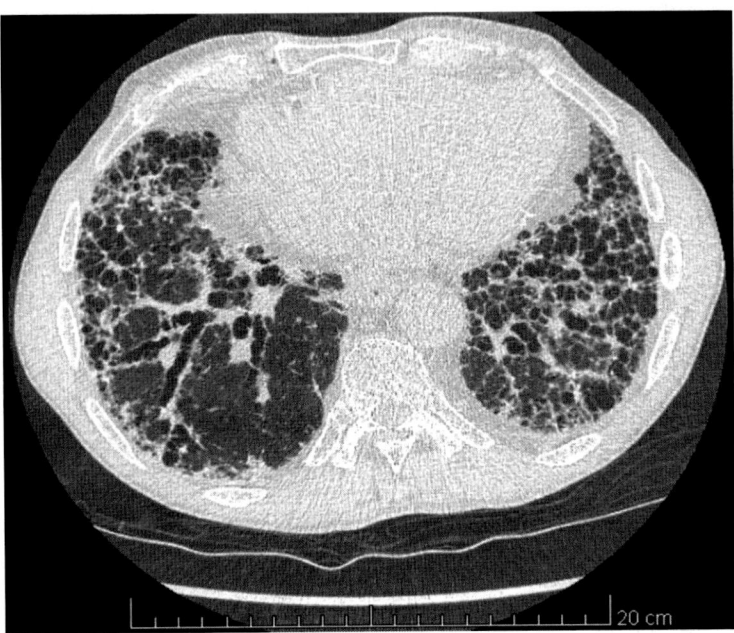

Fig. 3.17 Chest computed tomography (CT) scan showing the typical radiologic pattern of usual interstitial pneumonia (UIP), the hallmark for idiopathic pulmonary fibrosis, i.e., subpleural reticular opacities, honeycombing (clustered cystic air spaces), and traction bronchiectasis (dilatation of bronchi and bronchioles within areas of pulmonary fibrosis) with basal predominance (Lynch et al., 2018; Raghu et al., 2011). *(Image courtesy of Dr. Dan Adler.)*

Pache, Redard, & Gabbiani, 1995) has been demonstrated, supporting the participation of myofibroblasts in the evolution of these lesions. The origin of myofibroblasts is, as in practically all types of fibrosis, multiple; however, in the case of pulmonary fibrosis it appears that the main source is represented by local mesenchymal cells while epithelial cells and fibrocytes would serve as support through paracrine signaling (Chong, Sato, Kolb, & Gauldie, 2019; Hung, 2020).

The therapeutic armamentarium is so far limited to 2 drugs: nintedanib (an intracellular tyrosine kinase inhibitor; the tyrosine kinase playing a pivotal role in the activation of fibroblasts during fibrogenesis) and pirfenidone that can inhibit pro-fibrotic behaviors of fibroblasts and fibrocytes. These compounds may also prove clinically useful in reducing decline in lung function although their effect on symptoms is still questionable (Maher & Strek, 2019).

3.7 Liver fibrosis and cirrhosis

The liver has been an organ particularly important not only in the history of medicine, but also in that of human culture. Since Babylonian and Assyrian civilizations in the II millennium before Christ (BC), the operation of "inspecting the liver" in animals was considered fundamental before important decisions were taken. This because Babylonians thought that blood was the essence of life and liver the hematopoietic organ of the body, formed by coagulated blood. Therefore, the centrality of blood and liver characterized many ancient cultures. In classic medical system, firstly elaborated by Galen of Pergamon (also known as Pergamum or Pergamos), the liver continued to be responsible of blood production, being blood one of the four humors of the body with yellow bile, black bile, and phlegm. It was also the organ where two other humors were formed, the yellow bile and the black bile, stored in the gallbladder and in the spleen, respectively (Singer, 1957). The liver, finally, was the seat of production of the "natural spirit", the spirit in charge of sexual desire and vegetative functions, while the "vital spirit", produced in the left ventricle of the heart, was in charge of body heat and emotions, and the "animal spirit", produced in the brain, of the mind operations (Temkin, 1951). All along the history of ancient medicine, the fundamental organs of the body were indeed the liver, the heart and the brain, and the liver was considered the site of blood production until the 20th century.

Liver fibrosis and cirrhosis was characterized for the first time quite late in the history of medicine, despite the fact that one of its typical complications, the "ascites", was recognized since the antiquity. Ascites is the transliteration from the Greek term *askites*, which derives from the Homeric *áskos* that means water bag. *Askites*, which literally signifies "baglike", indicated dropsy of the peritoneum since the IV-III century BCE. A use confirmed by the Roman physicians Galen (Marcovecchio, 1993) and Aulus Cornelius Celsus (Jarcho, 1958). However, given that ascites can be a complication also of other conditions, such as cancer or heart failure, the history of this concept is not directly related to that of liver cirrhosis. Only sporadically, this association was made: for instance, the anatomist and physician Erasistratus from Alexandria associated ascites with liver disease in the III century BCE.

The first morphological description of liver cirrhosis appeared much later, dating back to Giovanni Battista Morgagni (1682–1771), Professor of Anatomy at the University of Padua between 1715 and 1771. In 1761, He published *De sedibus et causis morborum per anatomen indagatis* (On the Seats and Causes of Diseases as Investigated by Anatomy) in which he founded organ pathology (Morgagni, 1761). While in the previous centuries diseases were thought to be caused by an imbalance of humors, Morgagni demonstrated that clinical symptoms were systematically correlated with anatomical lesions in specific organs (Zampieri, Zanatta, & Thiene, 2014). His *De sedibus* was composed of 700 clinical cases with autopsy records in which the presence of a morphological substrate explaining physiopathology and symptomatology was demonstrated for any disease. The 700 cases were divided into five chapters, each one for a body part from head to toe, and 70 anatomo-medical letters, each one for a disease, a syndrome or a symptom. Morgagni is accounted for describing for the first time many diseases, among them the morphology of liver cirrhosis, even without recognizing the pathology. The term cirrhosis in fact did not exist at that time. In the anatomo-medical letter XXXVIII, "*On Ascites, Dropsy of the Peritoneum, and Hydatids*," Morgagni reported, in the paragraph 30th, the case of the Venetian patrician Gaspare Lombria who died in 1722 of a disease that now we can recognize as liver cirrhosis. Firstly, he visited the patient, diagnosing the presence of liquid in the bowels by a clinical operation still done today, the so-called "fluid thrill" or "fluid wave". He reported that "*Placing the left hand in a side of the abdomen and beating the other side with right hand, I could feel the flowing of the water from right to left inside the bowels*" (Morgagni, 1761, p.102).[o] At the autopsy, after discharging the water contained in the bowels, Morgagni observed "[...] *the liver hard, completely covered by deep and surface tubercles, that is, glandular little lobes very clearly distinguished*" (Morgagni, 1761, p. 103).[p] He gave the following interpretation: "*Surely the smallest parts of the liver could not increase to such an extent without consequences for the normal functioning of the organ and for the blood movement, by compressing other parts of the liver or at least the small vessels*" (Morgagni, 1761, p. 103).[q]

Morgagni did not relate this disease with alcohol consumption and explained the pathology assuming a hypertrophy of liver glands, which caused a slowing down of blood in the liver and in the bowels, which in turn caused the accumulation of water. Interestingly, he proposed the same explanation

---

[o] In Latin, in the original text "*Principio, si abdominis latus sinistra complectens, latus alterum repetitis dexterae modicis ictibus impelleres; allisae ad sinistram aquae fluctuationem percipiebas*" (Morgagni, 1761, p. 102).

[p] In Latin, in the original text "*Jecur durum, intus, extraque totum constans ex tuberculis,* id est *glandulosis lobulis evidetissimis, & evidentissime distinctis*" (Morgagni, 1761, p. 102).

[q] In Latin, in the original text "*Non possunt autem niminae jecinoris partes adeo amplificari, quin aut interjectas alias, aut vascula saltem sanguifera comprimendo, hepatis muneri, & sanguinis per ventrem, motui plurimum officiant*" (Morgagni, 1761, p. 103).

for a different case, similar to this one only for the ascites as clinical mani-
festation. In the same anatomo-medical letter, at the paragraph 28, in fact,
Morgagni reported that[r]:

> During the dissection of an old woman, died in Padua's Hospital toward the end of
> 1716 for a moderate ascites, we observed the following evidences. After cutting and
> discharging the water contained in the bowels, the liver appeared completely cov-
> ered by deep and surface nodules of whitish color and not very hard consistence. In
> the pancreas we find a similar one, but much harder and as big as to fill all its part
> connected to duodenum.

(Morgagni, 1761, p. 101)

   Morgagni compared this disease, which was probably a pancreas neopla-
sia with hepatic metastasis, with one observed in sheep, when he was young,
which was much probably an ascitogenic cirrhosis. This diagnostic error can
be explained by the fact that Morgagni advanced the same pathophysiological mechanism for these different affections, that is, a glandular hypertrophy
with small vessels compression, thus leading to the accumulation of fluid
in the peritoneal cavity. This model, according to which the body was a
glandular machine, derived from the iatromechanical school of Marcello
Malpighi (1628–1694), considered by Morgagni his spiritual master. The
blood, circulating through the vessels, was filtered by different glands specific
for each organ, thus producing the physiological humors: salivary glands
produced the saliva, testicles the sperm, liver the bile, kidneys the urine and
so on. Even the nervous system was considered to be filled by a nervous
fluid produced by glands supposed to existing in the brain (Bertoloni Meli,
2012). Therefore, most human diseases were related to the disorder of a specific set of glands. In Morgagni's *De sedibus* many diseases are explained with
this model, even liver cirrhosis, as discussed above (Fig. 3.18).
   René-Théophile-Marie-Hyacinthe Laennec (1781–1826) a French
physician, member of the Paris anatomic-clinical school revolutionized
medicine during the early XIX century by advancing Morgagni's method.
In particular, Laennec developed the anatomic-clinical approach with the
exploration of the living body in order to detect organic lesions related to
a disease. He advanced the method of auscultation inventing an instrument,

---

[r] In Latin, in the original text " *Vetulam quae ex hydrope ascite, sed non ita magno, decesserat, cum
in Nosocomio Patavino secaremus sub finem A. 1716 haec observavimus. Venter, ubi exhausta fuit
effusa aqua, jecur ostendit multis albis, nec tamen praeduris, tumoribus intus, extraque obsessum; in
pancreate autem similem unum, sed duriorem, multoque majorem, ut quod totam illam hujus visceris
occupabat partem qua ad Duodenum intestinum se annexit*" (Morgagni, 1761, p. 101).

THE PARACENTESIS.

Fig. 3.18 A surgeon performing a paracentesis on an obese man, whose swollen abdomen has a cannula inserted into it, and is subsequently releasing fluid into a basin. Pen drawing by Z.S. after an engraving, 1672. The original engraving is in: (Barbette, 1672) (p. 57). *(Credit: Wellcome Collection. Attribution 4.0 International (CC BY 4.0).)*

the stethoscope, which has become the symbol of medicine. Thanks to the stethoscope he described cardiac murmurs and discovered that tuberculosis was a disease always characterized by cavities ("tubercles") in the lungs. Before him, tuberculosis was called phthisis, that is, consumption, because the disease was understood only clinically. After him, the Paris anatomo–clinical school started a huge program in order to detect the specific morphological organ substrates affected by a given disease.

In 1819, Laennec published his famous *De l'auscultation médiate ou traité du diagnostic des maladies des poumons et du cœur fondé principalement sur ce nouveau moyen d'exploration* (Mediated auscultation or treatise on the diagnosis of lung and heart diseases based mainly on this new means of exploration) (Laënnec, 1819), also called *Le Traité* (The Treatise) in which he described his new invention and its applications, in particular on heart and lungs diseases. See Fig. 3.19.

However, this book is famous also because, describing the disease of the liver which most typically caused ascites, he coined the term "cirrhosis". See Fig. 3.20. First, Laennec described the macroscopic appearance of the organ with the following words[5]:

> "The liver, reduced to one third of its volume [...] had an external surface slightly corrugated and wrinkled, and a yellowish-gray color; cut, it seemed totally composed of a multitude of little round or ovoid grains [...]. Theses grains, easily separable each other, did not leave among them any space in which the liver normal tissue was still visible; their color was reddish or reddish-yellow, blending in some areas to greenish; their tissue, enough damp and opaque, to the touch was flaccid, rather than soft, and squashing the grains between the fingers, only a little part was pressed, while the rest gave to the touch the sensation of soft leather". At the end of

---

[5] The original text says (in French): "*Le foie, réduit au tiers de son volume ordinaire [...]; sa surface externe, légèrement mamelonnée et ridée, offrait une teinte grise-jaunâtre; incisé, il paraissait entièrement composé d'une multitude de petits grains de forme ronde ou ovoïde [...]. Ces grains, faciles à séparer les uns des autres, ne laissaient entre eux pres-qu'aucun intervalle dans lequel on pût distinguer encore quelque reste du tissu propre du foie; leur couleur était fauve ou d'un jaune roux, tirant par endroits sur le verdâtre; leur tissu, assez humide, opaque, était flasque au toucher plutôt que mou, et en pressant les grains entre les doigts, on n'en écrasait qu'une petite partie: le reste offrait au tact la sensation d'un morceau de cuir mou (a)*". The footnote (a) is: "*Cette espèce de production est encore du nombre de celles que on confond sous le nom de squirrhe. Je crois devoir la désigner sous le nom de cirrhose, à cause de sa couleur. Son développement dans le foie est une des causes les plus communes de l'ascite, et à cela de particulier qu'à mesure que les cirrhoses se développent, le tissu du foie est absorbé; qu'il finit souvent, comme chez ce sujet, par disparaître entièrement; et que, dans tous les cas, un foie qui contient des cirrhoses perd de son volume au lieu de s'accroître d'autant. Cette espèce de production se développe aussi dans d'autres organes, et finit par se ramollir comme toutes les productions morbidiques*". See Fig. 3.20.

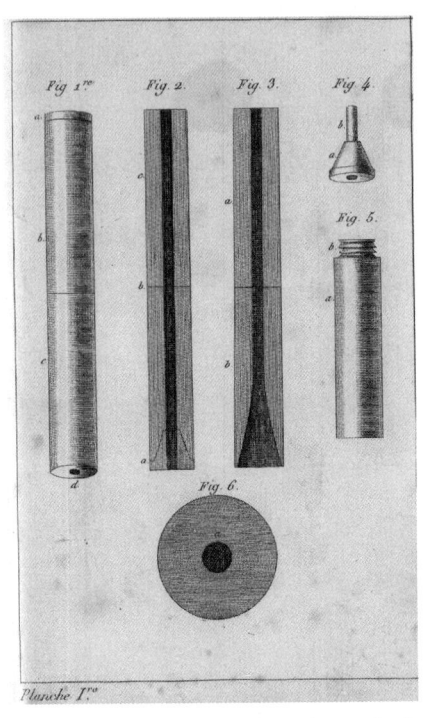

DE

L'AUSCULTATION

MÉDIATE

OU

TRAITÉ DU DIAGNOSTIC DES MALADIES

DES POUMONS ET DU CŒUR,

FONDÉ PRINCIPALEMENT SUR CE NOUVEAU
MOYEN D'EXPLORATION.

PAR R. T. H. LAENNEC,

D. M. P., Médecin de l'Hôpital Necker, Médecin honoraire
des Dispensaires, Membre de la Société de la Faculté de
Médecine de Paris et de plusieurs autres sociétés nationales
et étrangères.

Μέγα δὲ μέρος ἐγούμαι τῆς τέχνης εἶναι
τὸ δύνασθαι σκοπεῖν.
Pouvoir explorer est, à mon avis, une
grande partie de l'art. HIPP., *Épid. III.*

TOME PREMIER.

A PARIS,

CHEZ J.-A. BROSSON et J.-S. CHAUDÉ, Libraires,
rue Pierre-Sarrazin, n° 9.

1819.

Planche I.re

Fig. 3.19 On the left: the frontispiece of the first edition of Laennec's "*De l'auscultation mediate*" (1819), where he introduced for the first time the stethoscope. On the right, illustrations from the same book (Table 1, Fig. 1–6), of the first model of stethoscope. It is worth noticing that the Laennec's nomenclature is still in use nowadays (both in the French version, e.g., "*râles*", as well as in the English translation, e.g., "*rhonchi*"). Interestingly, Laennec used the term *rhonchus* (actually a synonym of *râle*, i.e., rattle) from the second edition of his *Traité*. The reason to choose the term *rhonchus* instead of *râle* could have been dictated by the fact of wanting not to frighten patient: *râle* being a term commonly used to describe the breathing sounds of a dying person (*râle de la mort*, death rattle) (da Silva, Tonietto, & Tonietto, 2017). According to the Byzantine Greek physician Paul of Aegina (Paulus Aegineta, Παῦλος Αἰγινήτης; c. 625-c. 690), the author of the 7th century CE "Medical Compendium in Seven Books" (in Latin: *Epitomae medicae libri septem,* in Greek: *Επιτομής Ιατρικής βιβλία επτά*) (Aegineta, 1844), the term *rhonchus* is probably a latinized form of the Greek term rhochmos, i.e., wheezing or snoring (ῥωχμός, from the verb rhocho, ῥώχω, to snore) (Aegineta, 1844) (Book III, p.482). This term had been previously used by the 1st century CE Greek physician Araeteus ('Ἀρεταῖος) of Cappadocia (written ῥοχμός) (Aretaeus, 1958) (1.11), by Aëtius of Amida ('Ἀέτιος 'Ἀμιδηνός), a byzantine Greek physician of mid-5th to mid-6th century CE (written ῥοχμός) (Medicus, 1534) (6.3), and by the Roman physician of the 5th century CE, Caelius Aurelianus (written ῥογμός and ῥογχός) (Aurelianus, 1709) (2.27, and 10 respectively). *(Credit: Wellcome Collection. Attribution 4.0 International (CC BY 4.0).)*

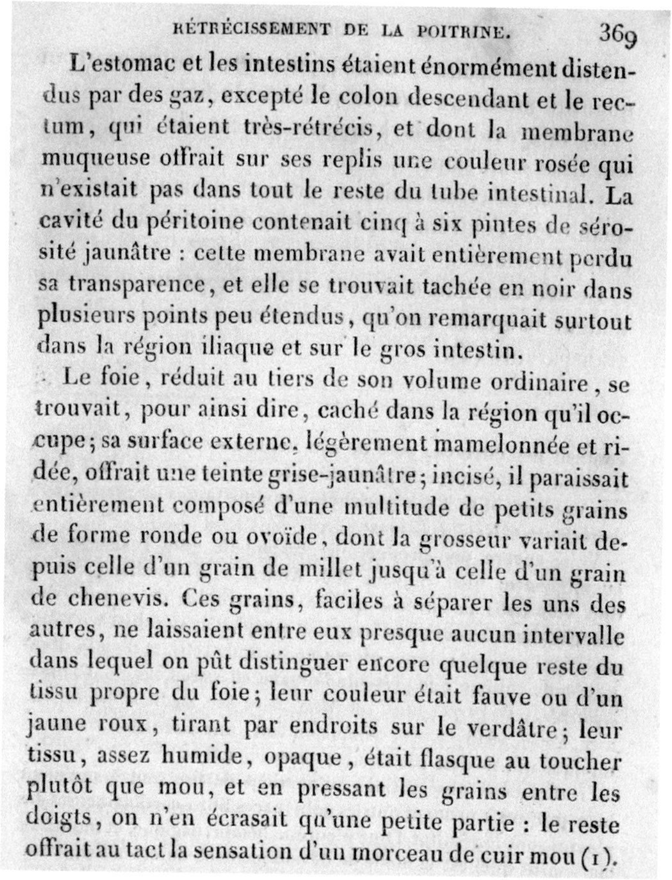

RÉTRÉCISSEMENT DE LA POITRINE.   369

L'estomac et les intestins étaient énormément disten-
dus par des gaz, excepté le colon descendant et le rec-
tum, qui étaient très-rétrécis, et dont la membrane
muqueuse offrait sur ses replis une couleur rosée qui
n'existait pas dans tout le reste du tube intestinal. La
cavité du péritoine contenait cinq à six pintes de séro-
sité jaunâtre : cette membrane avait entièrement perdu
sa transparence, et elle se trouvait tachée en noir dans
plusieurs points peu étendus, qu'on remarquait surtout
dans la région iliaque et sur le gros intestin.

Le foie, réduit au tiers de son volume ordinaire, se
trouvait, pour ainsi dire, caché dans la région qu'il oc-
cupe; sa surface externe, légèrement mamelonnée et ridée, offrait une teinte grise-jaunâtre; incisé, il paraissait
entièrement composé d'une multitude de petits grains
de forme ronde ou ovoïde, dont la grosseur variait de-
puis celle d'un grain de millet jusqu'à celle d'un grain
de chenevis. Ces grains, faciles à séparer les uns des
autres, ne laissaient entre eux presque aucun intervalle
dans lequel on pût distinguer encore quelque reste du
tissu propre du foie; leur couleur était fauve ou d'un
jaune roux, tirant par endroits sur le verdâtre; leur
tissu, assez humide, opaque, était flasque au toucher
plutôt que mou, et en pressant les grains entre les
doigts, on n'en écrasait qu'une petite partie : le reste
offrait au tact la sensation d'un morceau de cuir mou (1).

Fig. 3.20 Page 369 from Laennec's "De l'auscultation mediate", Tome 2 (Volume 2) where he describes the cirrhosis of the liver. In the note (a) he explains why he gave to this pathology the name cirrhosis: because of its yellowish color ("à cause de sa couleur"), since kirrhos (κιρρός) in Greek means tawny, yellowish/orangish-brown. *(Credit: Page 369 from Laennec's "De l'auscultation mediate", Tome 2 (Volume 2).)*

this phrase, Laennec wrote a footnote in which the term "cirrhosis" for the first time appeared: "[…] I think that this kind of morbid condition could be designed with the name cirrhosis because of its color. Its development in the liver is one of the most commons causes of ascites. This particularly because, as much the cirrhosis develop, as much the liver tissue is absorbed to the point of disappearing, as in this patient."

(Laënnec, 1819)

Laennec borrowed the name from the old Greek *kirrhos* (κιρρός), which meant indeed tawny, yellowish/orangish-brown. Laennec's description of liver cirrhosis was not limited to this one. In fact, he wrote, between 1804 and 1808, several hundred pages of a manuscript titled *Treatise of Pathological Anatomy* containing 16 pages dedicated to cirrhosis. Even if the treatise remained unpublished, it circulated among the European medical community, to the point to be quoted and discussed in some Medical Encyclopedias and Dictionaries (Duffin, 1987).

The English translation of Laennec's "*Traité*" (A Treatise of the Diseases of the Chest) was made by the Scottish physician John Forbes (1787–1861) (Forbes, 1821).

In this way, the eponym of Laennec's cirrhosis became history. Nevertheless, it should be noted that the disease had already been described, before, not only by Morgagni, but also by the British pathologists John Browne (1642—1700) and Matthew Baillie (1761–1823). During the XVIII century excessive consumption of cheap spirits produced an epidemic of cirrhosis, popularly known as "gin liver" in England and "brandy liver" elsewhere, which attracted also the attention of physicians. Although Browne and Baillie had used different names for the lesion, both had recorded its appearance accurately (Fig. 3.21).

Baillie, in particular, is an important historical figure for pathological anatomy because of his *The Morbid Anatomy of Some of the Most Important Parts of the Human Body* (Baillie, 1793). As already stated, organ pathology was founded by Morgagni, however Baillie's publication had a deep impact on the development of pathology as a separate subject of study, that would become later a medical specialty. This because the first edition, in

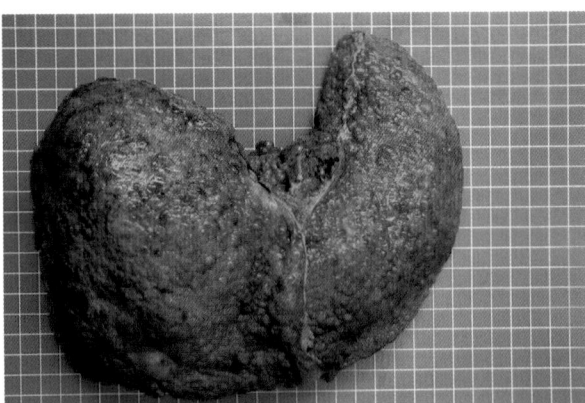

Fig. 3.21 Typical macroscopic aspect of alcoholic cirrhosis showing an irregular surface with "tubercles" as described originally by Morgagni. *(Image courtesy of Prof. Laura Rubbia-Brandt.)*

particular, was completely dedicated to the morphological description of diseased organs, with a clear and simple style, and without any digression on anatomical or clinical topics. Baillie was careful not to go beyond his observations, and his modest goal prevented him from analyzing, correlating, or theorizing about his data.

At the beginning of the book, Baillie stated that:

> *The natural structure of the different parts of the human body has been very minutely examined, so that anatomy may be said to have arrived at a high pitch of perfection; but our knowledge of the changes of structure produced by disease, which may be called the Morbid Anatomy, is still very imperfect.*
>
> **(Baillie, 1793, p. 34)**

Although explanations of morbid appearances have changed radically since his time, Baillie's descriptions are still of value. He was quite logical in his procedures, and he carefully distinguished between inflamed states, thickenings, hardenings, softening, ulcers, tubercles, aneurysms, and the like. Baillie's *Morbid Anatomy* had a wide diffusion, with eight English editions in rapid succession until 1838, three American editions from 1795 to 1820, two French (1803 and 1815), and one German (1793), Italian (1807) and Russian (1826) (Rodin 1973). Most subsequent pathology textbooks have been modeled on his, limited to the description of the structural changes produced by a series of well-known diseases, divided by tissues, organs and apparatuses. Since the second edition (1797) Baillie joined brief clinical considerations to the anatomo-pathological descriptions (Baillie, 1797).

Since the first edition of the *Morbid Anatomy*, Baillie linked liver cirrhosis to excessive alcohol consumption:

> *One of the most common diseases in the liver (and perhaps the most common, except the adhesions which we have lately described) is the formation of tubercles in its substance. This disease seldom affects a very young person, but frequently takes place in persons of middle or advanced age: it is likewise more common in men than in women. This would seem to depend upon the habit of drinking being more common in the one sex than in the other; for this disease is most frequently found in hard drinkers; although we cannot see any necessary connection between that mode of life and this particular disease in the liver.*
>
> **(Baillie, 1793, p. 136)**

Baillie proposed the correlation between alcohol consumption and liver cirrhosis, without any physio-pathological explanation. In the second edition, while describing clinical symptoms, Baillie stated to be unaware of specific symptoms related to the disease. However, according to him, when the patient felt pain in the area of the liver, had a yellowish color of the skin and an ascites, there were few doubts on the presence of a liver with tubercles (Baillie, 1797).

In the XIX century, new and more powerful microscopes allowed the British anatomist Francis Kiernan (1800–1874) to discover and describe, in 1833, the lobules of liver (Kiernan, 1833). As far as liver cirrhosis, the British Pathologist Sir Robert Carswell (1793–1857) was the first to histologically differentiate the disease. He was also skillful in drawing, and when he was still a medical student, was employed by Dr. John Thompson of Edinburgh to produce a collection of drawings illustrating morbid anatomy. In 1838, Carswell published a text-atlas entitled *Pathological Anatomy: Illustrations of the Elementary Forms of Disease* (Carswell, 1838), where he illustrated and described the histological features of liver cirrhosis. Carsweel stated that cirrhosis gave rise to dropsy and consisted in atrophy of the lobular structures of the organ *"produced by the presence of a contractile fibrous tissue formed in the capsule of Glisson"* (Carswell, 1838). (Description of Plate II) Carswell was convinced that the *"cellulo-fibrous or fibrous tissue now forms a conspicuous feature in the disease"* (Carswell, 1838) (Chapter "Atrophy"). In 1875, the Paris physician Victor Hanot (1844–1896) distinguished primary biliary cirrhosis from the other forms, but his work was not recognized until the German internist Carl von Liebermeister (1833–1901), clearly differentiated two forms of cirrhosis: portal (the classic Laennec's cirrhosis) and biliary (Franken, 1983).

In 1911, the American pathologist Frank Burr Mallory (1862–1941) introduced the entity of *"alcoholic hepatitis"*, identifying it as a precursor lesion of cirrhosis (Mallory, 1911). The Belgian born American naturalized clinical nutritionist Charles Saul Lieber (1931–2009) definitively demonstrated in 1968 that alcohol was directly hepatotoxic in humans and cirrhogenic in baboons, even in subjects who have an adequate diet, contradicting then-current scientific opinion which identified in different nutritional deficiencies the main causes of the disease (Hatroft, 1955).

The second half of the 20th century has seen important advances in the understanding of several aspects of hepatic fibrosis pathogenesis, clinical evolution, and therapy among which one of the most remarkable has been the identification of the cell now accepted as the mostly involved in the first phases of connective tissue accumulation and remodeling process during fibrosis development, i.e., the perisinusoidal stellate cell (PSC). PSCs, identified in the second half of the 19th century by Boll in 1869 (Boll, 1869) and more in detail by von Kupfer in 1876 (Von Kupfer, 1876), have been described extensively in the 20th century by several authors including Ito and Nemoto (Ito & Manji, 1952), Wake (Wake, 1971) and Aterman (Aterman, 1986) and have received several names such as Ito cells, stellate cells, fat storing cells as we shall see below, according to their morphology, function or the name of a scientist involved in their discovery.

PSCs are located in the space of Disse, just below endothelial cells and can be recognized by argentic impregnation, gold chloride staining (originally used by von Kupfer in his first report on these cells), by the presence of lipid droplets in their cytoplasm, hence the name of fat storing cells, or by the classical autofluorescence of vitamin A that is stored in lipid granules (Wake, 1971). Another feature of PCSs is the presence of cytoplasmic desmin expressing intermediate filaments; this feature is common in rats, but much less in humans (Schmitt-Gräff, Krüger, Bochard, Gabbiani, & Denk, 1991). The isolation and culture of PCSs has allowed establishing their capacity of producing collagens type I, III and IV at least in vitro (Friedman, Roll, Boyles, & Bissell, 1985). Several authors have suggested that PCSs could represent the precursor cell of collagen producing fibroblasts during the development of experimental or human fibrotic conditions (Kent et al., 1976; K.M. Mak & Lieber, 1988; W.W. Mak, Sattler, & Pitot, 1980; Wake, 1980). Moreover, cells with myofibroblastic features have been described in experimental and human liver fibrosis and cirrhosis (Irle, Kocher, & Gabbiani, 1980; Leo & Lieber, 1983; Leo, Mak, Savolainen, & Lieber, 1985; Rudolph, McClure, & Woodward, 1979).

A systematic study of normal and pathologic liver specimens has established that, while desmin positivity is rare in human PCSs compared to rat PCSs, α-SMA can be frequently expressed by these cells during development and pathological situations leading to fibrosis and cirrhosis (Fig. Xyz, from Schmitt-Gräff et al., 1991). These findings have been confirmed and now is well accepted that PCSs represent the main precursor of cirrhotic myofibroblasts (for review see Berumen, Baglieri, Kisseleva, & Mekeel, 2021; Friedman, 2008). This does not exclude that other cells, in particular portal fibroblasts, can also participate to the accumulation of a myofibroblastic population, e.g., during biliary cirrhosis (for review see Berumen et al., 2021; Hinz et al., 2012). The expression of α-SMA by PCSs can be reversible (Rubbia-Brandt et al., 1997) at least under certain pathological situations such as during acute hepatic ischemia: this indicates that α-SMA expression by these cells is not necessarily related to their modulation into myofibroblasts and may reflect other functional activities (Fig. 3.22).

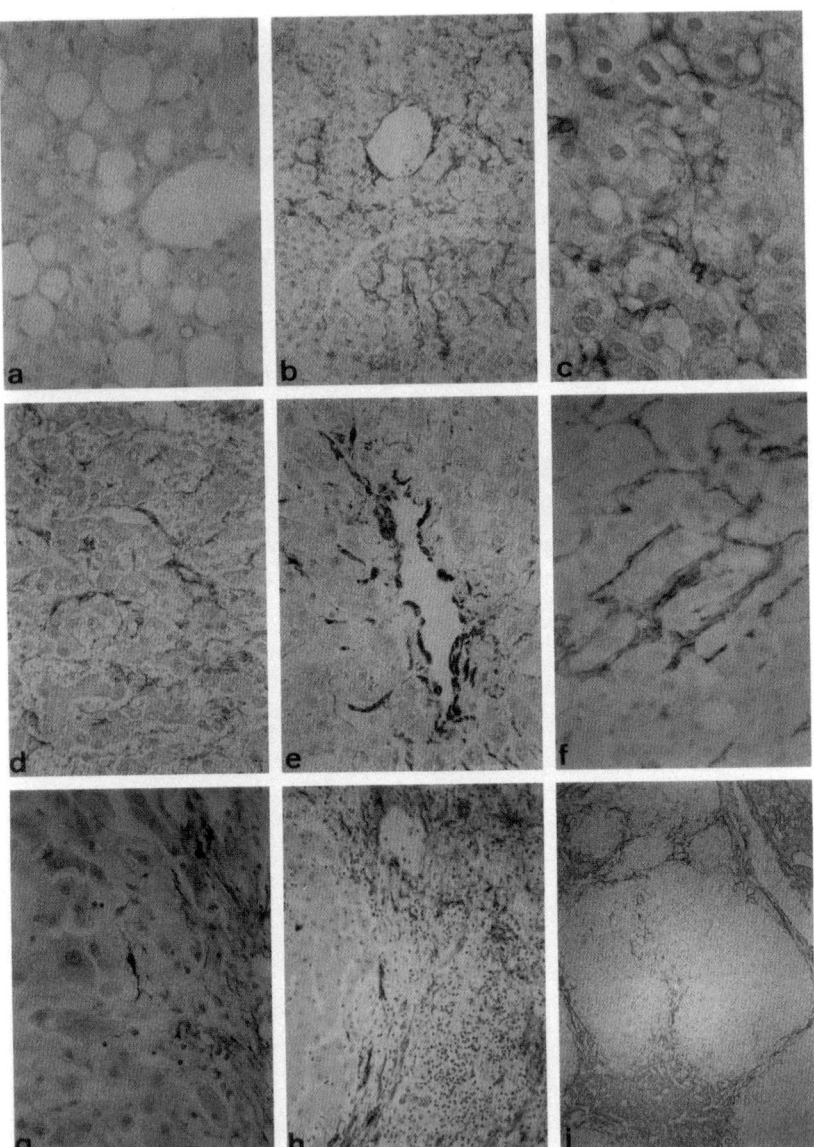

Fig. 3.22 Immunoperoxidase staining for α-SMA. In a fatty liver of a diabetic patient that does not show inflammatory infiltration, immunoreactive perisinusoidal (PCS) cells are absent (A, ×200). α-SMA containing PSCs are visible in a case of vitamin A intoxication characterized by a conspicuous proliferation of PSCs containing lipid vacuoles (B, ×100; C, ×550); in chronic congestion (D, ×200); in alcoholic liver disease with central sclerosis associated with many PSCs around a terminal hepatic venule (E, ×200); in alcoholic hepatitis with prominent pericellular fibrosis (F, ×450); in chronic active hepatitis with stellate-shaped PSCs at the margin of piecemeal necrosis (G, ×400); in chronic graft-versus-host disease with piecemeal necrosis containing PSCs in fibrous tissue near the portal tract (H, ×100); in active cirrhosis with alcoholic hepatitis a conspicuous immunoreactivity is visible in stromal cells of the scars, in fibrous septa expanding in the parenchyma, and in PSCs scattered within the nodules (I, ×60). *(Fig. 2 from Schmitt-Gräff, A., Krüger, S., Bochard, F., Gabbiani, G., & Denk, H. (1991). Modulation of α-SMA and desmin expression in perisinusoidal cells of normal and diseased human livers. The American Journal of Pathology, 138(5), 1233–1242. Copyright © 1991 Elsevier Masson SAS. All rights reserved.)*

3.8 Stroma reaction to epithelial tumors

Stroma Reaction (SR), a term utilized for the first time at the beginning of the last century (Russel, 1908), indicates an accumulation of inflammatory cells, fibroblasts and small vessels surrounding malignant cell nests of different tumors e.g., breast, colon, and pancreatic carcinoma (Frieberg, Nielsen, Mortensen, & Detlefsen, 2016), whose structure resembles that of a fibrotic tissue, such as granulation tissue and fibromatosis. For a long time, SR has been considered a response to tumor invasion without any significant biological activity. During the last years it has been gradually shown by many laboratories that SR, and particularly its fibroblastic cells, can influence the biology of epithelial cells and thus exert an important role in processes such as cancer invasion and metastasis formation (Liao, Tan, Zhu, & Tan, 2019). SR fibroblasts, also called cancer-associated fibroblasts (CAF), exhibit phenotypic features of myofibroblasts, including the expression of α-SMA (Cintorino et al., 1991), and have been shown to exert both beneficial and damaging influences on cancer development (Liao et al., 2019). Importantly, the presence of CAFs with myofibroblastic features has been shown to take place in the human uterine cervix before the occurrence of epithelial cell invasion, suggesting that CAFs may participate in cancer progression. See Fig. 3.23 from Cintorino et al. (1991).

Over the years a new theory on cancer development, the Organization Field Theory, has joined the classical Somatic Mutation Theory, suggesting that the microenvironment can exert a crucial action in cancer evolution (Liao et al., 2019). The mechanisms through which SR myofibroblasts may affect cancer evolution include mechanical force generation, production and secretion of several substances such as cytokines and growth factors, as well as of enzymes degrading extracellular matrix (Liao et al., 2019; Sanford-Crane, Abrego, & Sherman, 2019). An important feature of CAFs is their phenotypic heterogeneity that allegedly corresponds to a functional heterogeneity (Louault, Li, & DeClerck, 2020; Wang et al., 2021). Pinpointing this heterogeneity will be instrumental in organizing therapeutic strategies affecting the biology of stromal cells, aiming at influencing tumor progression, evolution and metastasis formation (Chen & Song, 2019).

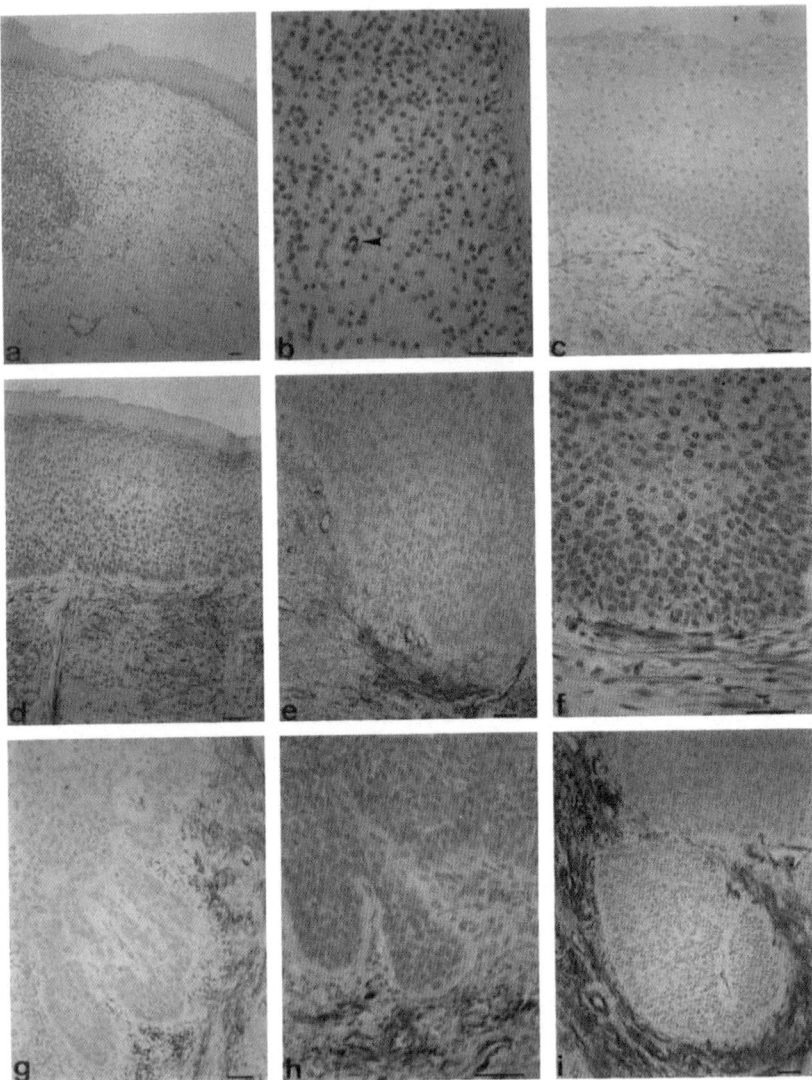

Fig. 3.23 Immunoperoxidase staining for α-SMA of sections from paraffin-embedded tissue blocks. Bars: 20 μm in a and 20 μm in b to i. (A) Low power view of a follicular cervicitis showing positive staining in several small vessels and no staining of stromal cells. (B) A higher magnification of the same lesion shows that anti-α-SMA antibody staining is confined to the wall of small vessels *(arrow)*. (C) CIN I with slight but clear positivity in several stromal cells. (D) CIN II with a stronger positivity in an important proportion of stromal cells. (E) CIN III showing abundant positive stromal cells, mainly adjacent to the basement membrane. (F) A detail of a CIN III lesion showing that the staining is localized in elongated stromal cells. (G) Low-power view of a micro-invasive carcinoma with a high proportion of anti-SMA -positive stromal cells. (H) High -power view of the same type of lesion showing several positive stromal cells adjacent to the invading nests of cells. (I) Invasive carcinoma with a typical desmoplastic reaction, with cells positive to anti-α-SMA. *(Fig. 1 from Cintorino, M., Bellizzi de Marco, E., Leoncini, P., Tripodi, S. A., Xu, L. J., Sappino, A. P., Schmitt-Gräff, A., & Gabbiani, G. (1991). Expression of a-smooth-muscle actin in stromal cells of the uterine cervix during epithlial neoplastic changes. International Journal of Cancer, 47(6), 843–846. Copyright © 1991 Elsevier Masson SAS. All rights reserved.)*

References

Abdelaziz Mohamed, I., Gadeau, A. P., Hasan, A., Abdulrahman, N., & Mraiche, F. (2019). Osteopontin: A promising therapeutic target in cardiac fibrosis. *Cell, 8*(12). https://doi.org/10.3390/cells8121558.

Adams, W. (1878). Contraction of the fingers (Dupuytren's contraction), and its successful treatment by subcutaneous divisions of the Palma fascia, and immediate extension. *British Medical Journal, 1*(913), 928–932. https://doi.org/10.1136/bmj.1.913.928.

Aegineta, P. (1844). *The seven books of Paulus Aegineta, translated from the Greek, with a commentary embracing a complete view of the knowledge possessed by the Greeks, Romans and Arabians on all subjects connected with medicine and surgery, by Francis Adams.* London: Syndenham Society of London.

Alexandre, P. (2005). Report on the Surgical Clinic at the Hotel-Dieu (original title in french: "Leçons orales de clinique chirurgicale faites à l'Hôtel-Dieu de Paris, par M. le Bon Dupuytren"). *Journal of Hand Surgery, 30*(6), 546–550.

Alibert, J.-L.-M. (1810). *Précis théorique et pratique sur les maladies de la peau.* Paris: chez le Dr. Daynac.

Allison, J. R., Jr., & Allison, J. R., Sr. (1966). Knuckle pads. *Archives of Dermatology, 93*(3), 311–316.

Aretaeus. (1958). *Aretaeus, edidit C. Hude.* Berlin Academiae Scientarum.

Aterman, K. (1986). The parasinusoidal cells of the liver: A historical account. *The Histochemical Journal, 18*(6), 279–305. https://doi.org/10.1007/bf01675207.

Aurelianus, C. (1709). *Celeres passiones.* Amsterdam: Amman.

Baillie, M. (1793). *The morbid anatomy of some of the most important parts of the human body.* London: J. Johnson, St. Paul's Church-yard, and G. Nicol, Pall-Mall, London.

Baillie, M. (1797). *The morbid anatomy of some of the most important parts of the human body.* London: J. Johnson & G. Nicol.

Barasch, M. (2000). The idol in the icon: Some ambiguities. In J. Assmann, & A. I. Baumgarten (Eds.), *Representation in religion: Studies in honor of Moshe Barasch* Leiden Brill.

Barbette, P. (1672). *Opera Chirurgico-Anatomica, ad circularem sanguinis motum, aliaque recentiorum inventa, accommodata. Accedit De Peste Tractatus, observationibus illustratus.* Lugd. Batav: Ex Officina Hackiana.

Barnett, A. (1974). *Scleroderma: Progressive systemic sclerosis.* Springfield, IL: Charles C Thomas.

Belusa, L., Selzer, A. M., & Partecke, B. D. (1995). Description of Dupuytren disease by the Basel physician and anatomist Felix Plater in 1614. *Handchirurgie, Mikrochirurgie, Plastische Chirurgie, 27*(5), 272–275.

Bertoloni Meli, D. (2012). *Mechanism, experiment, disease. Marcello Malpighi and seventeenth-century anatomy.* Baltimore: John Hopkins University Press.

Berumen, J., Baglieri, J., Kisseleva, T., & Mekeel, K. (2021). Liver fibrosis: Pathophysiology and clinical implications. *Wiley Interdisciplinary Reviews. Systems Biology and Medicine, 13*(1). https://doi.org/10.1002/wsbm.1499, e1499.

Boll, F. (1869). Die Bundessubstanz der Drusen. *Arch Mickr Ana, 5*, 334–355.

Boyer, A. (1831). Traité des maladies chirurgicales, et des opérations qui leur conviennent. *Paris: L'auteur [etc.].*

Bright, R. (1836). Cases and observations illustrative of renal disease, accompanied with the secretion of albuminous urine. *The Medico-Chirurgical Review, 25*(49), 23–35. Retrieved from https://pubmed.ncbi.nlm.nih.gov/29918407. https://www.ncbi.nlm.nih.gov/pmc/articles/PMC5093576/.

Buggey, J., Mentz, R. J., Pitt, B., Eisenstein, E. L., Anstrom, K. J., Velazquez, E. J., & O'Connor, C. M. (2015). A reappraisal of loop diuretic choice in heart failure patients. *American Heart Journal, 169*(3), 323–333. https://doi.org/10.1016/j.ahj.2014.12.009.

Capusan, I. (1972). Curzio's case of scleroderma. *Annals of Internal Medicine, 76*(1), 146. https://doi.org/10.7326/0003-4819-76-1-146_2.

Carswell, R. (1838). *Pathological anatomy: Illustrations of the elementary forms of disease.* Paternoster-row, London: Longman, Orme, Brown, Green, and Longman.

Chen, X., & Song, E. (2019). Turning foes to friends: Targeting cancer-associated fibroblasts. *Nature Reviews. Drug Discovery, 18*(2), 99–115. https://doi.org/10.1038/s41573-018-0004-1.

Chong, S. G., Sato, S., Kolb, M., & Gauldie, J. (2019). Fibrocytes and fibroblasts – where are we now. *The International Journal of Biochemistry & Cell Biology, 116,* 105595. https://doi.org/10.1016/j.biocel.2019.105595.

Cintorino, M., Bellizzi de Marco, E., Leoncini, P., Tripodi, S. A., Xu, L. J., Sappino, A. P., Schmitt-Gräff, A., & Gabbiani, G. (1991). Expression of α-smooth-muscle actin in stromal cells of the uterine cervix during epithlial neoplastic changes. *International Journal of Cancer, 47*(6), 843–846.

Cooper, A. (1851). *A treatise on dislocations and fractures of the joints.* Philadelphia: Blanchard and Lea.

Crunkhorn, S. (2009). Silencing microRNA rescues the heart. *Nature Reviews Drug Discovery, 8*(2), 109. https://doi.org/10.1038/nrd2810.

Curzio, C. (1753). *Discussioni anatomico-pratiche di un raro, e stravagante morbo cutaneo in una giovane donna felicemente curato in questo grande Ospedale degl'Incurabili indirizzate al chiarissimo Signor Abate Nollet, membro della Real Academia delle Scienze in Parigi, e Maestro di Fisica del Serenissimo Delfino.* Napoli: Giovanni di Simone.

da Silva, V. D., Tonietto, R. G., & Tonietto, V. (2017). Laennec, R.T.H. (1781–1826). In J. G. van den Tweel (Ed.), *Pioneers in pathology* (pp. 313–321). Cham: Springer International Publishing.

de Lapeyronie, F. (1743). Sur quelques obstacles s'opposant à l'éjaculation naturelle de la semence. *Mem. Acad. Roy. de chir., I, 1743,* 318.

Delaunay, M., Osman, H., Kaiser, S., & Diviani, D. (2019). The role of cyclic AMP signaling in cardiac fibrosis. *Cell, 9*(1), 69. https://doi.org/10.3390/cells9010069.

Devine, C. J., Jr., & Horton, C. E. (1988). Peyronie's disease. *Clinics in Plastic Surgery, 15*(3), 405–409.

Dinarès Solà, R., Baxarias, J., Fontaine, V., Garcia-Guixé, E., & Herrerín, J. (2012). Estudio radiológico realizado a 18 momias egipcias a pie de tumba. *Imagen Diagnóstica, 3*(1), 11–23. https://doi.org/10.1016/S2171-3669(12)70044-6.

Duff, G. L. (1948). The diffuse collagen diseases; a morphological correlation. *Canadian Medical Association Journal, 58*(4), 317–325.

Duffin, J. M. (1987). Why does cirrhosis belong to Laennec? *Canadian Medical Association Journal, 137*(5), 393–396.

Dupuytren, G. (1831). De la retraction des doigts par suite d'une affection de l'aponevrose palmaire. - Descripition de la maladie. - Operation chirurgicale qui convient dans ce cas. Compte rendu de la clinique chirurgicale de l'Hôtel Dieu par MM. les docteurs Alexandre Paillard et Marx. *Journal Universel et Hebdomadaire de Medicine et de Chirurgie Pratiques et des Institutions Medicales, 5,* 127–141.

Dupuytren, G. (1832). Retraction permanente des doigts. In *Vol. I. Leçons de clinique chirurgicale faites a l'Hotel Dieu de Paris* (pp. 2–24). Paris: Chez Germer Baillière.

Dupuytren, G. (1834). Permanent retraction of the fingers produced by an affection of the palmar fascia. *Lancet, 2,* 222–225. Retrieved from https://ci.nii.ac.jp/naid/10005047517/en/.

Durel, L. (1888). Essai sur la maladie de Dupuytren. Thèse de doct. de Paris, n° 227.

Ehrlich, H. P., Desmoulière, A., Diegelmann, R. F., Cohen, I. K., Compton, C. C., Garner, W. L., … Gabbiani, G. (1994). Morphological and immunochemical differences between keloid and hypertrophic scar. *The American Journal of Pathology, 145*(1), 105–113.

Elliot, D. (1988). The early history of contracture of the palmar fascia. Part 1: The origin of the disease: The curse of the MacCrimmons: The hand of benediction: Cline's contracture. *The Journal of Hand Surgery, 13*(3), 246–253. https://doi.org/10.1016/0266-7681(88)90078-2.

Evans, J. N., Kelley, J., Low, R. B., & Adler, K. B. (1982). Increased contractility of isolated lung parenchyma in an animal model of pulmonary fibrosis induced by bleomycin. *The American Review of Respiratory Disease, 125*(1), 89–94. https://doi.org/10.1164/arrd.1982.125.1.89.

Evans, R. D., & Nuzum, F. R. (1929). Fibrosis of the myocardium. *California and Western Medicine, 30*, 11–16.

Falloppia, G. (1606). *Opera genuina omnia, tam practica quam theorica.. quorum pars una, tota praesertim chirurgia et tractatus de morbo gallico, methodusque consultandi ab auctore ad editionem concinnata.. : nunc primum lucem adspicit, pars vero altera e volumine incondito Francofurti nuper edito desumpta.. et.. repurgata.* Venetiis: J. A. et J. de Franciscis.

Fang, L., Murphy, A. J., & Dart, A. M. (2017). A clinical perspective of anti-fibrotic therapies for cardiovascular disease. *Frontiers in Pharmacology, 8*. https://doi.org/10.3389/fphar.2017.00186.

Fantonetti, G. (1837). Skleroderma generale. *Effemeridi delle Scienze Mediche, 1*, 41–45.

Fierro-Fernández, M., Miguel, V., Márquez-Expósito, L., Nuevo-Tapioles, C., Herrero, J. I., Blanco-Ruiz, E., … Lamas, S. (2020). MiR-9-5p protects from kidney fibrosis by metabolic reprogramming. *The FASEB Journal, 34*(1), 410–431. https://doi.org/10.1096/fj.201901599RR.

Flatt, A. E. (2001). The Vikings and Baron Dupuytren's disease. *Proceedings (Baylor University. Medical Center), 14*(4), 378–384. https://doi.org/10.1080/08998280.2001.11927791.

Forbes, J. (1821). *A treatise of the diseases of the chest.* London: Underwood.

Forget, C. (1847). Memoire sur le chorionitis ou la sclerostenose cutanee (maladie non decrit par les auteurs). *Gaz Med Strasbourg, 7*(200–12).

Franken, F. (1983). *History of hepatology.* Berlin, Heidelberg: Springer-Verlag.

Frieberg, M., Nielsen, B., Mortensen, M. B., & Detlefsen, S. (2016). Key players in pancreatic cancer-stroma interaction: Cancer-associated fibroblasts, endothelial and inflammatory cells. *World Journal of Gastroenterology, 22*, 2678–2700.

Friedman, S. L. (2008). Hepatic stellate cells: Protean, multifunctional, and enigmatic cells of the liver. *Physiological Reviews, 88*(1), 125–172. https://doi.org/10.1152/physrev.00013.2007.

Friedman, S. L., Roll, F. J., Boyles, J., & Bissell, D. M. (1985). Hepatic lipocytes: the principal collagen-producing cells of normal rat liver. *Proceedings of the National Academy of Sciences of the United States of America, 82*, 8681–8685.

Futterman, B. (2015). Analysis of the Papal Benediction sign: The ulnar neuropathy of St. Peter. *Clinical Anatomy, 28*(6), 696–701. https://doi.org/10.1002/ca.22584.

Gabbiani, G., & Majno, G. (1972). Dupuytren's contracture: fibroblast contraction? An ultrastructural study. *The American Journal of Pathology, 66*(1), 131–146.

Gabrielli, A., Avvedimento, E. V., & Krieg, T. (2009). Scleroderma. *The New England Journal of Medicine, 360*(19), 1989–2003. https://doi.org/10.1056/NEJMra0806188.

Gallizia, F. (1964). Triade collagene: maladie de la Peyronie, maladie de Dupuytren et fibrose du cartilage auriculaire. *Journal of Urology & Nephrolog (Paris), 70*, 424–427.

Garrod, A. (1893). On an unusual form of nodule upon the joints of the fingers. *St. Bartholomew's Hospital Reports, 29*, 157–161.

Garrod, A. E. (1904). Concerning pads upon the finger joints and their clinical relationships. *British Medical Journal, 2*(2270), 8. https://doi.org/10.1136/bmj.2.2270.8.

Ghazawi, F. M., Zargham, R., Gilardino, M. S., Sasseville, D., & Jafarian, F. (2018). Insights into the pathophysiology of hypertrophic scars and keloids: How do they differ? *Advances in Skin & Wound Care, 31*(1), 582–595. https://doi.org/10.1097/01.ASW.0000527576.27489.0f.

Gintrac, E. (1847). Note sur la sclérodermie. *Revue médico-chirurgicale, 2*, 263–267.

Goetz, R. (1945). Pathology of progressive systemic sclerosis (generalized scleroderma) with special reference to changes in the viscera. *Clinical Proceedings (South Africa), 4*, 337–342.

Grant, R. L. (1960). Antyllus and his medical works. *Bulletin of the History of Medicine, 34*(2), 154–174. Retrieved from http://www.jstor.org/stable/44446677.

Green, R. (1951). *A translation of Galen's hygiene (De Sanitate Tuenda)*. Springfield, Ill: Charles C. Thomas.

Gullberg, D., Kletsas, D., & Pihlajaniemi, T. (2016). Editorial: Wound healing and fibrosis—Two sides of the same coin. *Cell and Tissue Research, 365*(3), 449–451. https://doi.org/10.1007/s00441-016-2478-7.

Hamman, L. R., & Rich, A. R. (1944). Acute diffuse interstitial fibrosis of the lungs. *Bulletin of the Johns Hopkins Hospital, 74*, 177–212.

Hatroft, W. (1955). General histopathology and histopathology of the liver. *Journal of the National Cancer Institute, 15*, 1463–1468.

Henderson, N. C., Mackinnon, A. C., Farnworth, S. L., Kipari, T., Haslett, C., Iredale, J. P., … Sethi, T. (2008). Galectin-3 expression and secretion links macrophages to the promotion of renal fibrosis. *The American Journal of Pathology, 172*(2), 288–298. https://doi.org/10.2353/ajpath.2008.070726.

Hinz, B., Phan, S. H., Thannickal, V. J., Prunotto, M., Desmoulière, A., Varga, J., … Gabbiani, G. (2012). Recent developments in myofibroblast biology: Paradigms for connective tissue remodeling. *The American Journal of Pathology, 180*(4), 1340–1355. https://doi.org/10.1016/j.ajpath.2012.02.004.

Hort, W. (2002). History of cardiovascular pathology. *Zeitschrift für Kardiologie, 91*, iv20–iv24.

Horteloup, P. (1865). *De la sclérodermie*. Paris: P. Asselin.

Hoyne, G. F., Elliott, H., Mutsaers, S. E., & Prêle, C. M. (2017). Idiopathic pulmonary fibrosis and a role for autoimmunity. *Immunology and Cell Biology, 95*(7), 577–583. https://doi.org/10.1038/icb.2017.22.

Hughes, T. B., Mechrefe, A., Littler, J. W., & Akelman, E. (2003). Dupuytren's disease. *Journal of the American Society for Surgery of the Hand, 3*(1), 27–40. https://doi.org/10.1053/jssh.2003.50005.

Hung, C. (2020). Origin of myofibroblasts in lung fibrosis. *Current Tissue Microenvironment Reports, 1*(4), 155–162. https://doi.org/10.1007/s43152-020-00022-9.

Hutchison, R. L., & Rayan, G. M. (2011). Astley Cooper: His life and surgical contributions. *The Journal of Hand Surgery, 36*(2), 316–320. https://doi.org/10.1016/j.jhsa.2010.10.036.

Irle, C., Kocher, O., & Gabbiani, G. (1980). Contractility of myofibroblasts during experimental liver cirrhosis. *Journal of Submicroscopic Cytology, 12*, 209–217.

Ito, T., & Manji, N. (1952). Über Kupferschen Sternzellen und die "Fettspeicherungszellen" ("fat-storing cells") in der Blutkapillarenwand der menschlichen Leber. *Okajimas Folia Anatomica Japonica, 24*, 243–258.

Jarcho, S. (1958). Ascites as described by Aulus Cornelius Celsus (ca. A. D. 30). *The American Journal of Cardiology, 2*(4), 507–508. https://doi.org/10.1016/0002-9149(58)90339-4.

Kapanci, Y., Desmouliere, A., Pache, J. C., Redard, M., & Gabbiani, G. (1995). Cytoskeletal protein modulation in pulmonary alveolar myofibroblasts during idiopathic pulmonary fibrosis. Possible role of transforming growth factor beta and tumor necrosis factor alpha. *American Journal of Respiratory and Critical Care Medicine, 152*(6 Pt 1), 2163–2169. https://doi.org/10.1164/ajrccm.152.6.8520791.

Kaposi, M. (1881). *Leçon sur les maladies de la peau*. E. Besnier & A. Doyon, Trans. Paris: Masson.

Kent, G., Gay, S., Inouye, T., Bahu, R., Minick, O. T., & Popper, H. (1976). Vitamin A-containing lipocytes and formation of type III collagen in liver injury. *Proceedings of the National Academy of Sciences of the United States of America, 73*, 3719–3722.

Kiernan, F. (1833). XXIX. The anatomy and physiology of the liver. *Philosophical Transactions of the Royal Society, 123*, 711–770.

Klemperer, P. (1950). The concept of collagen diseases. *The American Journal of Pathology, 26*(4), 505–519.

Klemperer, P., Pollack, A. D., & Baehr, G. (1984). Landmark article May 23, 1942: Diffuse collagen disease. Acute disseminated lupus erythematosus and diffuse scleroderma. By Paul Klemperer, Abou D. Pollack and George Baehr. *JAMA*, *251*(12), 1593–1594. https://doi.org/10.1001/jama.251.12.1593.

Klinkhammer, B. M., Goldschmeding, R., Floege, J., & Boor, P. (2017). Treatment of renal fibrosis-turning challenges into opportunities. *Advances in Chronic Kidney Disease*, *24*(2), 117–129. https://doi.org/10.1053/j.ackd.2016.11.002.

Kodama, B. F., Gentry, R. H., & Fitzpatrick, J. E. (1993). Papules and plaques over the joint spaces. Knuckle pads (heloderma). *Archives of Dermatology*, *129*(8), 1044–1045. 1047 https://doi.org/10.1001/archderm.129.8.1044.

Kuppe, C., Ibrahim, M. M., Kranz, J., Zhang, X., Ziegler, S., Perales-Patón, J., … Kramann, R. (2021). Decoding myofibroblast origins in human kidney fibrosis. *Nature*, *589*(7841), 281–286. https://doi.org/10.1038/s41586-020-2941-1.

Laënnec, R. (1819). *De l'auscultation médiate ou traité du diagnostic de maladies des poumons et du coeur, fondé principalement sur ce nouveau moyen d'exploration*. Paris: Brosson et Chaudé.

Langhans, T. (1887). Histologische Teil im Rahmen der Arbeit Kocher TH, Die Behandlung der Retraktion der Palmaraponeurose. *Zentralblatt für Chirurgie*, *14*, 481–497.

Larsen, R. (1966). Dupuytren's contracture. Hand surgery. In J. Flynn (Ed.), *Hand surgery* (pp. 922–952). Baltimore: The William and Wilkins Co.

Ledderhose, G. (1894). Über zerreisungen der plantarfascie. *Archiv fur Klinische Chirurgie*, *48*, 853–856.

Leo, M. A., & Lieber, C. S. (1983). Hepatic fibrosis after long-term administration of ethanol and moderate vitamin A supplementation in the rat. *Hepatology*, *3*, 1–11.

Leo, M. A., Mak, K. M., Savolainen, E. R., & Lieber, C. S. (1985). Isolation and culture of myofibroblasts from rat liver. *Proceedings of the Society for Experimental Biology and Medicine*, *180*, 382–391.

Leroy, E. C. (1972). Connective tissue synthesis by scleroderma skin fibroblasts in cell culture. *The Journal of Experimental Medicine*, *135*(6), 1351–1362. https://doi.org/10.1084/jem.135.6.1351.

LeRoy, E. C., Black, C., Fleischmajer, R., Jablonska, S., Krieg, T., Medsger, T. A., Jr., … Wollheim, F. (1988). Scleroderma (systemic sclerosis): Classification, subsets and pathogenesis. *The Journal of Rheumatology*, *15*(2), 202–205.

Letulle, M. (1879). *Recherches sur les hypertrophies cardiaques secondaires*. Paris: Asselin.

Letulle, M. D. M. (1880). *Recherches anatomiques et cliniques sur l'hypertrophie cardiaque de la néphrite interstitielle*. Paris: Asselin.

Li, S., Yang, B., Du, Y., Lin, Y., Liu, J., Huang, S., … Zhang, Y. (2018). Targeting PPARα for the treatment and understanding of cardiovascular diseases. *Cellular Physiology and Biochemistry*, *51*(6), 2760–2775. https://doi.org/10.1159/000495969.

Liao, Z., Tan, Z. W., Zhu, P., & Tan, N. S. (2019). Cancer-associated fibroblasts in tumor microenvironment − Accomplices in tumor malignancy. *Cellular Immunology*, *343*, 103729. https://doi.org/10.1016/j.cellimm.2017.12.003.

Lipson, K. E., Wong, C., Teng, Y., & Spong, S. (2012). CTGF is a central mediator of tissue remodeling and fibrosis and its inhibition can reverse the process of fibrosis. *Fibrogenesis & Tissue Repair*, *5*(1), S24. https://doi.org/10.1186/1755-1536-5-S1-S24.

Louault, K., Li, R.-R., & DeClerck, Y. A. (2020). Cancer-associated fibroblasts: Understanding their heterogeneity. *Cancers*, *12*(11), 3108. https://doi.org/10.3390/cancers12113108.

Luck, J. V. (1959). Dupuytren's contracture; a new concept of the pathogenesis correlated with surgical management. *The Journal of Bone and Joint Surgery*, *41*(4), 635–664.

Lynch, D. A., Sverzellati, N., Travis, W. D., Brown, K. K., Colby, T. V., Galvin, J. R., … Wells, A. U. (2018). Diagnostic criteria for idiopathic pulmonary fibrosis: A Fleischner Society White Paper. *The Lancet Respiratory Medicine*, *6*(2), 138–153. https://doi.org/10.1016/s2213-2600(17)30433-2.

Mac Callum, J. (1889). A contribution to the knowledge of the pathology of fragmentation and segmentation, and fibrosis of the myocardium. *The Journal of Experimental Medicine, 4,* 409–424.

Madelung, O. (1875). Die Atiologie und die operative Behandlung der Dupuytren'schen Fingerverkriimmung. *Berliner Klinische Wochenschrift, 12*(191).

Maher, T. M., & Strek, M. E. (2019). Antifibrotic therapy for idiopathic pulmonary fibrosis: Time to treat. *Respiratory Research, 20*(1), 205. https://doi.org/10.1186/s12931-019-1161-4.

Mahon, H. E. M. (1966). Malignant nephrosclerosis—50 years later. *Journal of the History of Medicine and Allied Sciences, XXI*(2), 125–146. https://doi.org/10.1093/jhmas/XXI.2.125.

Mak, K. M., & Lieber, C. S. (1988). Lipocytes and transitional cells in alcoholic liver disease: A morphometric study. *Hepatology, 8,* 1027–1033.

Mak, W. W., Sattler, C. A., & Pitot, H. C. (1980). Accumulation of actin microfilaments in adult rat hepatocytes cultured on collagen gel/nylon mesh. *Cancer Research, 40,* 4552–4564.

Mallory, F. (1911). Cirrhosis of the liver. Five different types of lesions from which it may arise. *Johns Hopkins Hospital Bulletin, 22,* 69–75.

Marcovecchio, E. (1993). *Dizionario etimologico storico dei termini medici* (p. 1993). Impruneta-Firenze: Festina Lente.

McFarlane, R. M. (2002). On the origin and spread of Dupuytren's disease. *The Journal of Hand Surgery, 27*(3), 385–390. https://doi.org/10.1053/jhsu.2002.32334.

Medicus, A. (1534). *Venice: Editio Aldina.*

Meyrier, A. (2015). Nephrosclerosis: A term in quest of a disease. *Nephron, 129*(4), 276–282. https://doi.org/10.1159/000381195.

Mitchell, J., Woodcock-Mitchell, J., Reynolds, S., Low, R., Leslie, K., Adler, K., … Skalli, O. (1989). Alpha-smooth muscle actin in parenchymal cells of bleomycin-injured rat lung. *Laboratory Investigation, 60*(5), 643–650.

Moore-Morris, T., Guimarães-Camboa, N., Banerjee, I., Zambon, A. C., Kisseleva, T., Velayoudon, A., … Evans, S. M. (2014). Resident fibroblast lineages mediate pressure overload-induced cardiac fibrosis. *The Journal of Clinical Investigation, 124*(7), 2921–2934. https://doi.org/10.1172/jci74783.

Morgagni, G. (1761). *De sedibus et causis morborum per anatomen indagatis.* Venetiis: Typographia Remondiniana.

Musitelli, S., Bossi, M., & Jallous, H. (2008). Sexuam medicine history: A brief historical survey of "Peyronie's disease". *The Journal of Sexual Medicine, 5*(7), 1737–1746. https://doi.org/10.1111/j.1743-6109.2007.00692.x.

Nemeth, J., Schundner, A., & Frick, M. (2020). Insights into development and progression of idiopathic pulmonary fibrosis from single cell RNA studies. *Frontiers in Medicine, 7*(945). https://doi.org/10.3389/fmed.2020.611728.

Ng, M., Lawson, D. J., Winney, B., & Furniss, D. (2020). Is Dupuytren's disease really a 'disease of the Vikings'? *The Journal of Hand Surgery European, 45*(3), 273–279. https://doi.org/10.1177/1753193419882351.

Nicolle, M. (1890). *Contribution à l'étude des affections du myocarde. Les grandes scléroses cardiaques.* Paris: G. Steinheil.

Noble, P. W., & Homer, R. J. (2005). Back to the future: Historical perspective on the pathogenesis of idiopathic pulmonary fibrosis. *American Journal of Respiratory Cell and Molecular Biology, 33*(2), 113–120. https://doi.org/10.1165/rcmb.F301.

Osterman, A. M., Murray, P. M., & Pianta, T. (2012). Cline's contracture: Dupuytren was a thief – A history of surgery for Dupuytren's contracture. In *Dupuytren's disease and related hyperproliferative disorders principles, research, and clinical perspectives* (pp. 195–206). Berlin, Heidelberg: Springer-Verlag.

Papadakis, M., Manios, A., & Trompoukis, C. (2017). First report of palmar fibromatosis and camptodactyly in the 2nd century AD? *Acta Chirurgica Belgica, 117*(1), 64–66. https://doi.org/10.1080/00015458.2016.1261492.

Pasero, G., & Marson, P. (2004). Hippocrates and rheumatology. *Clinical and Experimental Rheumatology, 22*(6), 687–689.

Peitzman, S. J. (1989). From dropsy to Bright's disease to end-stage renal disease. *The Milbank Quarterly, 67*(Suppl 1), 16–32.

Petit, A. (2016). Histoire de la chéloïde [History of keloid]. *Annales de Dermatologie et de Vénéréologie, 143*, 81–95.

Platter, F. (1614). *Observationum, in Hominis Affectibus plerisq[ue], corpori & animo, functionum laesione, dolore, aliave molestia & vitio incommodantibus.* Basileae: Impensis Ludovici König, typis Conradi Waldkirchii.

Raghu, G., Collard, H. R., Egan, J. J., Martinez, F. J., Behr, J., Brown, K. K., … Schünemann, H. J. (2011). An official ATS/ERS/JRS/ALAT statement: Idiopathic pulmonary fibrosis: Evidence-based guidelines for diagnosis and management. *American Journal of Respiratory and Critical Care Medicine, 183*(6), 788–824. https://doi.org/10.1164/rccm.2009-040GL.

Rajkumar, V. S., Howell, K., Csiszar, K., Denton, C. P., Black, C. M., & Abraham, D. J. (2005). Shared expression of phenotypic markers in systemic sclerosis indicates a convergence of pericytes and fibroblasts to a myofibroblast lineage in fibrosis. *Arthritis Research & Therapy, 7*(5), R1113–R1123. https://doi.org/10.1186/ar1790.

Retz, N. (1790). *Des maladies de la peau et de celles de l'esprit (telles que les vapeurs, la mélancolie, la manie etc.), qui procèdent des affections du foie. Leur origine, la description de celles qui sont le moins connues, les traitements qui leur conviennent* (3e ed., pp. 155–158). Paris: chez Méquignon, l'aîné.

Richeldi, L., Collard, H. R., & Jones, M. G. (2017). Idiopathic pulmonary fibrosis. *The Lancet, 389*(10082), 1941–1952. https://doi.org/10.1016/s0140-6736(17)30866-8.

Rindfleisch, G. E. (1897). Über cirrhosis cystica pulmonum. *Verh. Ges. dtsch. Naturforsch., 8*, 864.

Rodnan, G. P., & Benedek, T. G. (1962). An historical account of the study of progressive systemic sclerosis (diffuse scleroderma). *Annals of Internal Medicine, 57*, 305–319. https://doi.org/10.7326/0003-4819-57-2-305.

Rolski, F., & Błyszczuk, P. (2020). Complexity of TNF-α signaling in heart disease. *Journal of Clinical Medicine, 9*(10), 3267. https://doi.org/10.3390/jcm9103267.

Ronchese, F. (1966). Knuckle pads and similar-looking disorders. *Giornale Italiano di Dermatolotia. Minerva Dermatologica, 107*(5), 1227–1235.

Rubbia-Brandt, L., Mentha, G., Desmoulière, A., Alto Costa, A. M., Giostra, E., Molas, G., … Gabbiani, G. (1997). Hepatic stellate cells reversibly express alpha-smooth muscle actin during acute hepatic ischemia. *Transplantation Proceedings, 29*, 2390–2395.

Rudolph, R., McClure, W. J., & Woodward, M. (1979). Contractile fibroblasts in chronic alcoholic cirrhosis. *Gastroenterology, 76*, 704–709.

Ruiz-Ortega, M., Rayego-Mateos, S., Lamas, S., Ortiz, A., & Rodrigues-Diez, R. R. (2020). Targeting the progression of chronic kidney disease. *Nature Reviews Nephrology, 16*(5), 269–288. https://doi.org/10.1038/s41581-019-0248-y.

Russel, B. R. (1908). The nature of the resistance to the inoculation of cancer. *Third Scientific Report of the Imperial Cancer Research Found, 3*, 341–358.

Sanford-Crane, H., Abrego, J., & Sherman, M. H. (2019). Fibroblasts as modulators of local and systemic cancer metabolism. *Cancers (Basel), 11*(5). https://doi.org/10.3390/cancers11050619.

Santucci, M., Borgognoni, L., Reali, U. M., & Gabbiani, G. (2001). Keloids and hypertrophic scars of Caucasians show distinctive morphologic and immunophenotypic profiles. *Virchows Archiv, 438*(5), 457–463. https://doi.org/10.1007/s004280000335.

Sappino, A. P., Masouyé, I., Saurat, J. H., & Gabbiani, G. (1990). Smooth muscle differentiation in scleroderma fibroblastic cells. *The American Journal of Pathology, 137*(3), 585–591.

Schafer, S., Viswanathan, S., Widjaja, A. A., Lim, W. W., Moreno-Moral, A., DeLaughter, D. M., … Cook, S. A. (2017). IL-11 is a crucial determinant of cardiovascular fibrosis. *Nature, 552*(7683), 110–115. https://doi.org/10.1038/nature24676.

Schenck von Grafenberg, J. (1596). *Observationvm medicarvm, rararvm, novarvm, admirabilivm, et monstrosarvm, liber quintus, de partibus externis; in qvo, qvae medici .. abdita, vulgo incognita, gravia, periculosaque, in harum partium conformationibus, earundemq; morborum causis, signis, eventibus, & curationibus accidere compererunt, exemplis utplurimum & historiis proposita exhibentur .. Continentur pr terea illustrium tatis nostr medicorum .. complura et nunquam publicata exempla memorabilia .. /.* Friburgi Brisgoiae: Ex officina Martini Bockleri.

Schmitt-Gräff, A., Krüger, S., Bochard, F., Gabbiani, G., & Denk, H. (1991). Modulation of alpha smooth muscle actin and desmin expression in perisinusoidal cells of normal and diseased human livers. *The American Journal of Pathology, 138*(5), 1233–1242.

Shearer, F., Lang, C. C., & Struthers, A. D. (2013). Renin-angiotensin-aldosterone system inhibitors in heart failure. *Clinical Pharmacology and Therapeutics, 94*(4), 459–467. https://doi.org/10.1038/clpt.2013.135.

Siegel, R. (1968). *Galen's system of physiology and medicine.* Basel-New York: Karger.

Singer, C. (1957). *A Short History of Anatomy and Physiology from the Greeks to Harvey.* New York: Dover Publications.

Skoog, T. (1948). Dupuytren's contraction: With special reference to etiology and improved surgical treatment: Its occurrence in knuckle pads. *Acta Chirurgica Scandinavica, 96*, 1–190.

Sobernheim, J. (1837). *Praktische Diagnostik der inneren Krankheiten mit vorzueglicher Ruecksicht auf pathologische Anatomie.* Berlin: Hirschwald.

Somers, K. D., & Dawson, D. M. (1997). Fibrin deposition in Peyronie's disease plaque. *The Journal of Urology, 157*(1), 311–315.

Stevenson, D. n.d. The Internet Classics Archive | Aphorisms by Hippocrates. http://classics.mit.edu/Hippocrates/aphorisms.html.

Sun, Y. B., Qu, X., Caruana, G., & Li, J. (2016). The origin of renal fibroblasts/myofibroblasts and the signals that trigger fibrosis. *Differentiation, 92*(3), 102–107. https://doi.org/10.1016/j.diff.2016.05.008.

Talman, V., & Ruskoaho, H. (2016). Cardiac fibrosis in myocardial infarction-from repair and remodeling to regeneration. *Cell and Tissue Research, 365*(3), 563–581. https://doi.org/10.1007/s00441-016-2431-9.

Temkin, O. (1951). On Galen's pneumatology. *Gesnerus, 8*, 180–189.

Tomasek, J. J., Gabbiani, G., Hinz, B., Chaponnier, C., & Brown, R. A. (2002). Myofibroblasts and mechano-regulation of connective tissue remodelling. *Nature Reviews. Molecular Cell Biology, 3*(5), 349–363. https://doi.org/10.1038/nrm809.

Volhard, F., & Fahr, T. (1914). *Die Brightsche Nierenkrankheit. Klinik, Pathologie und Atlas.* Berlin-Heidelberg: Springer-Verlag.

Von Buhl, L. (1872). *Lungenentzündung.* Munich Oldenburg: Tuberkulose und Schwindsucht.

Von Kupfer, C. (1876). Über Sternzellen der Lieber. Briefliche Mittheilung an Professor Waldeier. *Archiv für Mikroskopische Anatomie, 12*, 353–358.

Vrebos, J. (2009). G. Dupuytren's contracture: An inaccurate denomination. *Acta Chirurgica Belgica, 109*(5), 657–667. https://doi.org/10.1080/00015458.2009.11680512.

Vyalov, S. L., Gabbiani, G., & Kapanci, Y. (1993). Rat alveolar myofibroblasts acquire alpha-smooth muscle actin expression during bleomycin-induced pulmonary fibrosis. *The American Journal of Pathology, 143*(6), 1754–1765. Retrieved from https://pubmed.ncbi.nlm.nih.gov/7504890. https://www.ncbi.nlm.nih.gov/pmc/articles/PMC1887256/.

Wake, K. (1971). "Sternzellen" in the liver: Perisinusoidal cells with special reference to storage of vitamin A. *The American Journal of Anatomy, 132*(4), 429–462. https://doi.org/10.1002/aja.1001320404.

Wake, K. (1980). Perisinusoidal stellate cells (fat-storing cells, interstitial cells, lipocytes), their related structure in and around the liver sinusoids, and vitamin A-storing cells in extrahepatic organs. *International Review of Cytology, 66*, 303–353.

Wang, Z., Yang, Q., Tan, Y., Tang, Y., Ye, J., Yuan, B., & Yu, W. (2021). Cancer-associated fibroblasts suppress cancer development: The other side of the coin. *Frontiers in Cell and Developmental Biology*, *9*(146). https://doi.org/10.3389/fcell.2021.613534.

Webber, M., Jackson, S. P., Moon, J. C., & Captur, G. (2020). Myocardial fibrosis in heart failure: Anti-fibrotic therapies and the role of cardiovascular magnetic resonance in drug trials. *Cardiology and Therapy*, *9*(2), 363–376. https://doi.org/10.1007/s40119-020-00199-y.

Weber, K. T., Sun, Y., Bhattacharya, S. K., Ahokas, R. A., & Gerling, I. C. (2013). Myofibroblast-mediated mechanisms of pathological remodelling of the heart. *Nature Reviews. Cardiology*, *10*(1), 15–26. https://doi.org/10.1038/nrcardio.2012.158.

Wilson, E. (1847). On diseases of the skin. In *Diseases of the skin* (pp. 306–354). London: John Churchill.

Winterbauer, R. H. (1964). Multiple telangiectasia, Raynaud's phenomenon, sclerodactyly, and subcutanious calcinosis: A syndrome mimicking hereditary hemorrhagic telangiectasia. *Bulletin of the Johns Hopkins Hospital*, *114*, 361–383.

Yanagihara, T., Sato, S., Upagupta, C., & Kolb, M. (2019). What have we learned from basic science studies on idiopathic pulmonary fibrosis? *European Respiratory Review*, *28*(153), 190029. https://doi.org/10.1183/16000617.0029-2019.

Zampieri, F. (2016). *Il metodo anatomo-clinico tra meccanicismo ed empirismo: Marcello Malpighi, Antonio Maria Valsalva, Giovanni Battista Morgagni*. Roma: "L'Erma" di Bretschneider.

Zampieri, F., Zanatta, A., & Thiene, G. (2014). An etymological "autopsy" of Morgagni's title: De sedibus et causis morborum per anatomen indagatis (1761). *Human Pathology*, *45*, 12–16.

Zdilla, M. J. (2017). The hand of Sabazios: Evidence of Dupuytren's disease in antiquity and the origin of the hand of benediction. *The Journal of Hand Surgery (Asian-Pacific Volume)*, *22*(3), 403–410. https://doi.org/10.1142/s0218810417970012.

Zhang, A. Y., & Kargel, J. S. (2018). The basic science of Dupuytren disease. *Hand Clinics*, *34*(3), 301–305. https://doi.org/10.1016/j.hcl.2018.03.001.

Zhang, W. B., Du, Q. J., Li, H., Sun, A. J., Qiu, Z. H., Wu, C. N., … Ge, J. B. (2012). The therapeutic effect of rosuvastatin on cardiac remodelling from hypertrophy to fibrosis during the end-stage hypertension in rats. *Journal of Cellular and Molecular Medicine*, *16*(9), 2227–2237. https://doi.org/10.1111/j.1582-4934.2012.01536.x.

Ziegler, E. (1882). *Lehrbuch der allgemeinen und speciellen pathologischen Anatomie und Pathogenese*. Jena: Fischer.

CHAPTER 4

The myofibroblast: Role in fibrosis development

4.1 Discovery

The involvement of the myofibroblast in wound healing and fibrotic diseases is presently well accepted and has allowed a new approach to the study of these phenomena. As in many other cases, the discovery of the myofibroblast arose from an unexpected observation that is summarized in the following lines. At the end of the 1960s, one of us (GG), after receiving his PhD degree at the Institute of Experimental Medicine and Surgery of the University of Montreal under the direction of Prof. Hans Selye, was named Assistant Professor in the same institution with the charge of opening a laboratory of electron microscopy in order to continue his research work and eventually help other staff members with research problems involving this technique. GG obtained a fellowship of the Canadian Medical Research Council allowing a sabbatical year in the laboratory of 3 scientists: Guido Majno at the Department of Pathology, Harvard Medical School, Boston, United States, W. Stanley Hartroft, Hospital for Sick Children, University of Toronto, Canada and C.A Baud, Department of Anatomy, University of Geneva Medical School, Geneva, Switzerland. Working with G. Majno, GG became familiar with the new concept of nonmuscle cell contraction, in particular endothelial cell contraction, the main topic of Majno's laboratory. In Dr. Hartroft laboratory GG was supposed to study by means of electron microscopy the pyogenic membrane of an experimental sterile abscess in the rat in order to follow the formation of different ceroid pigments in macrophages: during this work he noticed that the granulation tissue fibroblasts, very numerous in the wall, were loaded with filaments, thus assuming, in a large proportion of cases, a morphology similar to that of smooth muscle cells. This observation immediately suggested that the filamentous structures could be implicated in fibroblast-related contractile phenomena playing a role in wound contraction and fibrotic retractions. It should be noted that at that time wound contraction was mainly related to collagen

Wound Healing, Fibrosis, and the Myofibroblast
https://doi.org/10.1016/B978-0-323-90546-6.00010-1

rearrangement, although in previous years some authors, e.g., Alexis Carrel or Michael Abercrombie, had suggested that this phenomenon was cell-dependent. Excited by the observation, GG tried to convince Dr. Hartroft and even the Chairman of the Department of Pathology Dr. A.Z. Movat, without success. On the contrary, when shortly after he visited Hans Selye in Montreal and showed him his observation he received an enthusiastic encouragement; indeed, Selye stated that such an observation could be the basis for a lifelong work. Similarly, Guido Majno was very positive and proposed to GG of move with him to Geneva, where he had just been offered the Chair of the Department of Pathology, in order to set up together a collaboration on the study of this newly discovered fibroblastic cell. GG accepted the proposal and moved with his family to Geneva in 1969, where he started to work with Majno and G.B. Ryan on this cell. When Majno and Ryan decided to go back respectively to United States and Australia, GG continued to develop this topic through his life as anticipated by his teacher.

The first description of the myofibroblast, based on the presence of microfilament bundles in fibroblasts from several experimental granulation tissues was published in Experientia after the manuscript had been refused by Nature (Fig. 4.1 from Gabbiani, Ryan, and Majno (1971)).

Soon thereafter gap junctions connecting myofibroblasts among them and focal adhesions connecting myofibroblasts with the extracellular matrix were identified (Gabbiani, Chaponnier, & Hüttner, 1978; Gabbiani, Hirschel, Ryan, Statkov, & Majno, 1972). Moreover, thanks to the excellent work of a PhD student, Omar Skalli (now professor at the University of Memphis), who produced a specific antibody, it was shown that myofibroblasts express α-SMA, the actin isoform typical of vascular SMCs (Darby, Skalli, & Gabbiani, 1990; Skalli et al., 1986). See Fig. 4.2 from Darby et al. (1990).

Moreover, it was shown that strips of granulation tissue from experimental animals and humans contract and relax in a pharmacological bath when treated with substances that contract or relax SMCs (Gabbiani et al., 1972). Gradually many features of myofibroblasts were described, such as their disappearance by apoptosis during the transition between granulation tissue and scar (Desmoulière, Redard, Darby, & Gabbiani, 1995) and their presence in hypertrophic scars (Santucci, Borgognoni, Reali, & Gabbiani, 2001) and in practically all fibrotic diseases including fibromatoses (Gabbiani & Majno, 1972) and scleroderma (Sappino, Masouyé, Saurat, & Gabbiani, 1990). Their presence in the stroma reaction of epithelial tumors

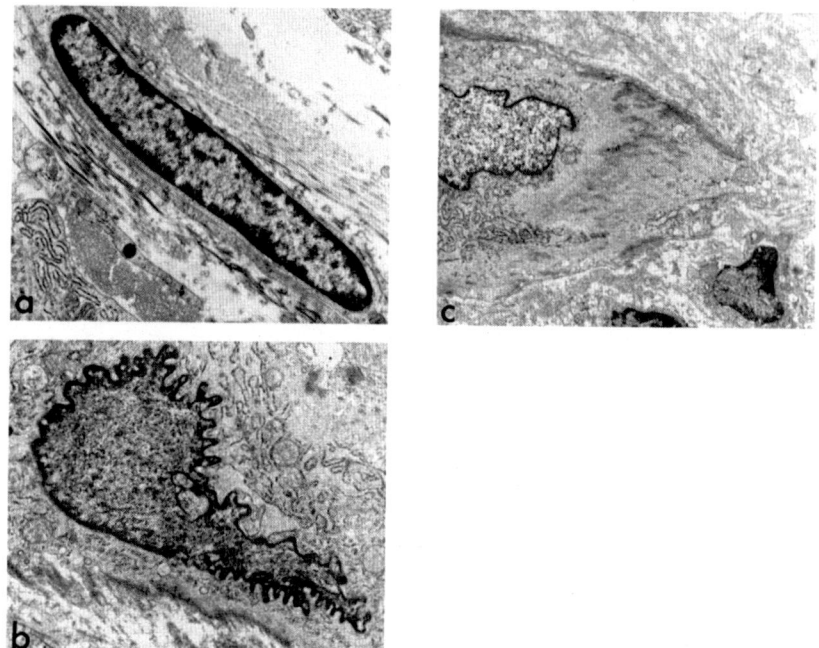

Fig. 4.1 Fibroblasts from control and granulation tissue. (A) Fibroblast from normal subcutaneous tissue: the nucleus is oval and the cell is surrounded by collagen fibers. ×10,500. (B) Modified fibroblast from the wall of a 21-day-old granuloma pouch: numerous folds and indentations of the nucleus. In addition to a prominent endoplasmic reticulum, the cytoplasm contains many fibrils located mostly at the periphery of the cell. In the extracellular space are collagen fibers with a finer fibrillar material without periodicity. ×14,000. (C) Another modified fibroblast showing a well-developed fibrillar system with dense bodies both within the cytoplasm and at the periphery of the cell, in connection with an extracellular layer of basement membrane-like material parallel to the cellular surface. ×7600. *(Fig. 1 from Gabbiani, G., Ryan, G. B., & Majno, G. (1971). Presence of modified fibroblasts in granulation tissue and their possible role in wound contraction.* Experientia, 27(5), 549–550. https://doi.org/10.1007/bf02147594. *Copyright ©*

was also reported (Liao, Tan, Zhu, & Tan, 2019). Moreover, the presence of myofibroblasts in normal organs of adult animals was described, supporting the possibility that they exert a physiological mechanical function (Kapanci, Assimacopoulos, Irle, Zwahlen, & Gabbiani, 1974). This possibility has been reinforced by the recent description of bona fide myofibroblasts in human and rat fascia tissues, lumbar fascia in particular, where it has been proposed that they play a role in active musculoskeletal dynamics (Schleip et al., 2019).

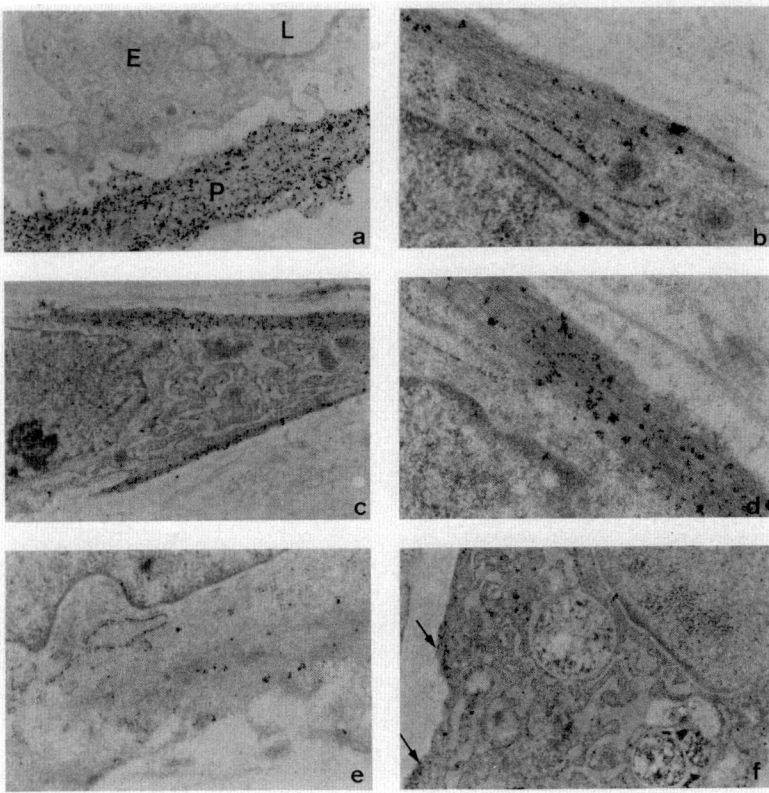

Fig. 4.2 Immunohistochemical localization of α-SMA in cells of granulation tissue. (A), section of 4-day-old wound tissue stained with anti-α-SMA antibody showing strong labeling of a pericyte (P) surrounding the endothelial cell € of a small vessel. L, vessel lumen. ×9400. (B) Eight-day-old wound tissue fibroblastic cell showing diffuse, although not strong, staining of the microfilaments present. ×43,270. (C) Twelve-day-old wound tissue myofibroblast shows prominent microfilament bundles strongly positive to anti-α-SMA. ×10,450. (D) Higher magnification of a 12-day-old wound tissue myofibroblast showing that α-SMA staining is limited to the microfilamentous bundles. ×33,430. (E) Twenty-day-old wound tissue fibroblast showing some microfilaments which are weakly stained with anti-α-SMA antibody. 27,750. (F) Twenty-five day wound tissue fibroblast showing early degenerative changes and weak staining at the periphery *(arrows)* with anti-α-SMA antibody. ×14,625. *(Fig. 5 from Darby, I., Skalli, O., & Gabbiani, G. (1990). Alpha-smooth muscle actin is transiently expressed by myofibroblasts during experimental wound healing. Laboratory Investigation, 63(1), 21–29. Copyright © 1990 Springer Nature. All rights reserved.)*

4.2 Origin

As expected, it was first suggested that myofibroblasts originate from local fibroblasts (Gabbiani et al., 1972). This remains true, possibly at least in part, for most fibrotic situations and particularly in the case of wound healing (Tomasek, Gabbiani, Hinz, Chaponnier, & Brown, 2002); however surprisingly, it was soon described that myofibroblasts can develop from several sources such as: (1) local connective tissue cells, i.e., pericytes, endothelial cells (through endothelial–mesenchymal transition) and perisinusoidal cells of the liver, (2) epithelial cells, through epithelial–mesenchymal transition and (3) bone marrow derived circulating cells called fibrocytes (Bochaton-Piallat, Gabbiani, & Hinz, 2016). The origin of myofibroblasts possibly varies according to the pathological situation and needs to be thoroughly investigated. A recent work using genetic lineage-tracing shows that after experimental myocardial infarction, granulation tissue myofibroblasts derive essentially from fibroblasts surrounding the area of ischemia (Tomasek et al., 2002). The variety of myofibroblast origins suggests that myofibroblast function is essential in several normal and pathological circumstances as well as that the myofibroblast represents a functional state of several cells rather than a single cell type.

The observation that a nonmuscle cell was exerting contractile activities and was equipped with contractile proteins integrated well with the then emerging concept of cytoskeleton, an apparatus present in all cells that was related to maintenance of cell shape and cell and organelle movement (Pollard, 1976). It was rapidly recognized that nonmuscle cells are equipped with different fibrillar elements, e.g., microfilaments, intermediate filaments and microtubules that exert static and dynamic functions and regulate several important biological functions [for a review of the development of the cytoskeleton concept see (Zampieri, Coen, & Gabbiani, 2014)]; now the cytoskeleton has become a chapter in Cell Biology textbooks. See Fig. 4.3 from Hinz et al. (2012).

4.3 Biological features

Initially it was assumed that myofibroblasts contract similarly to SMCs, however it was soon realized that force production by these cells is rather depending from the activation of the Rho/ROCK/myosin light chain phosphatase pathway (Tomasek et al., 2002). Moreover, myofibroblasts use mon muscle myosin rather than muscle myosin (Tomasek et al., 2002). This type of slow contraction corresponds, rather than to the classical muscle

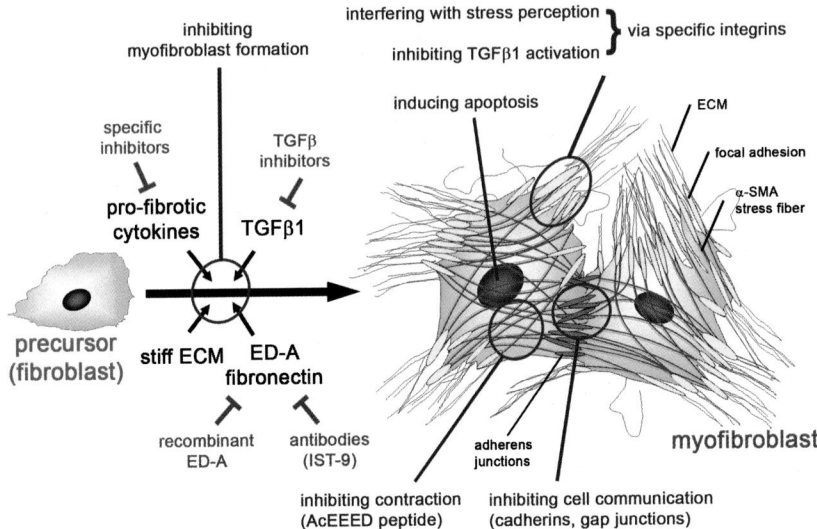

Fig. 4.3 Antifibrotic therapies can be designed to interfere with the extracellular chemical and mechanical factors that lead to the myofibroblast formation from a variety of different precursors. Alternatively, one might interfere with intracellular signaling pathways, transcription regulators, and epigenetic mechanisms that specifically modulate myofibroblast differentiation. Other potential antifibrotic targets are specific features of the differentiated myofibroblast, such as α-SMA in the contractile apparatus, specific integrins, and ECM proteins. *(Fig. 3 from Hinz, B., Phan, S. H., Thannickal, V. J., Prunotto, M., Desmoulière, A., Varga, J., ... Gabbiani, G. (2012). Recent developments in myofibroblast biology: Paradigms for connective tissue remodeling.* The American Journal of Pathology, *180(4), 1340–1355. Copyright © 2012 Springer Nature. All rights reserved.)*

contraction mechanism, to a gradual tissue remodeling activity. The force generated by the myofibroblast it transmitted to the extracellular matrix through focal adhesions containing integrins (Bochaton-Piallat et al., 2016). The overall result is a strained and compacted extracellular matrix, which in turn can stimulate myofibroblast force production.

The regulation of myofibroblast development concentrates a growing experimental attention and allegedly will produce important practical results in the near future. Recently the production by myofibroblasts of substances, such as hyaluronodase-2 (Midgley et al., 2020) or proteoglycan-4 (Qadri et al., 2020), which can influence positively or negatively their functions, as well as substances, such as phosphoinositide 3-kinase3, which can influence the behavior of other cells, such as cancer epithelial cells (Gagliano et al., 2020) and thus condition the evolution of different physiological and pathological processes, has been reported.

4.4 Heterogeneity

Myofibroblasts may possess different phenotypic features, at least in part in relation with their different origin. One can distinguish proto-myofibroblasts expressing only cytoplasmic actin isoforms with nonmuscle myosin in their stress fibers, well-differentiated myofibroblasts expressing α-SMA and in some instances myofibroblasts expressing muscle myosin (Kapanci et al., 1974). See Fig. 4.4.

Fibroblast heterogeneity has been studied in different situations, including stroma reaction to epithelial cancers. Moreover, fibroblast subgroups with different biological features, corresponding to different promoting capacities, start to be identifier and defined (Kang et al., 2021).

We do not know if different myofibroblast phenotypes produce different soluble substances paying a role in fibrosis evolution, but this aspect is actively investigated by several laboratories and new data should be available in the near future (Gagliano et al., 2020).

4.5 Factors influencing myofibroblast biology

The evolution of fibrosis is allegedly regulated by composition and organization of the extracellular matrix (Herrera, Henke, & Bitterman, 2018). The early modulation from fibroblast into proto-myofibroblasts allegedly depends on the increased stiffness of the extracellular matrix, which takes place at the beginning of granulation tissue formation, following tissue injury. Under the exposure to increased stiffening, fibroblasts develop stress fibers that contain only cytoplasmic actins: at this time, they are called proto-myofibroblasts. This change is generally followed by fibronectin accumulation and the splice variant of cellular fibronectin EDA is necessary to the activity of TGF-β1 in stimulating the transition between proto-myofibroblast and differentiated α-SMA expressing myofibroblast and are called differentiated myofibroblasts (Serini et al., 1998; Tomasek et al., 2002). See Fig. 4.5.

A recent review article on the role of fibronectin in development and wound healing summarizes well the participation of cellular fibronectin in normal and pathological wound healing (Patten & Wang, 2021). α-SMA expression by the differentiated myofibroblast increases the contractility of this cell and thus promotes tissue remodeling (Fu et al., 2018). Thus, it appears that mechanical forces, such as tissue stiffening are instrumental for the development of the typical myofibroblast biological features (Hinz,

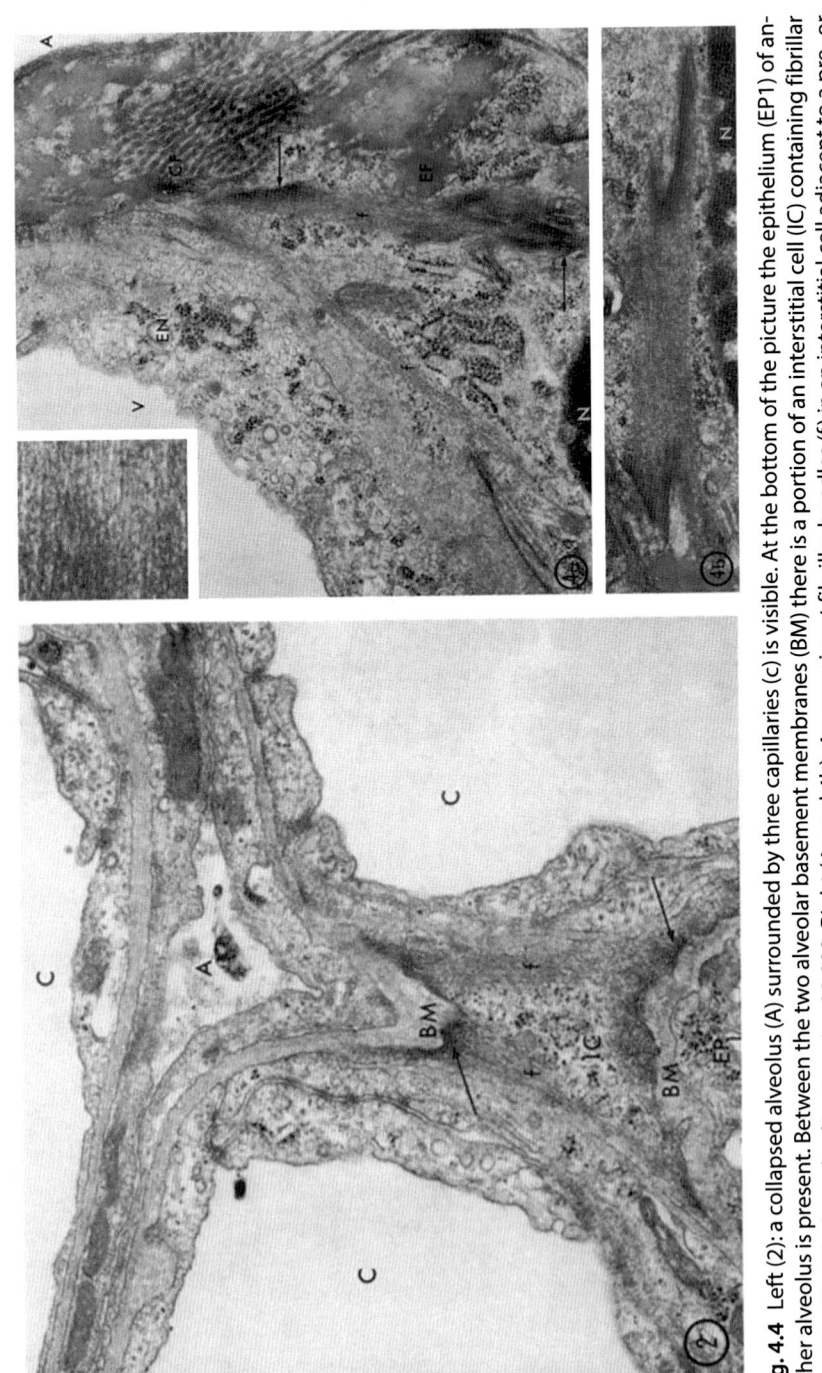

Fig. 4.4 Left (2): a collapsed alveolus (A) surrounded by three capillaries (c) is visible. At the bottom of the picture the epithelium (EP1) of another alveolus is visible. Between the two alveolar basement membranes (BM) there is a portion of an interstitial cell (IC) containing fibrillar bundles (f) with dense bodies (f). Right (4a and 4b): 4a, prominent fibrillar bundles (f) in an interstitial cell adjacent to a pre- or postcapillary vessel (V). Note the dense bodies (*arrows*) appearing as if they were anchoring the cell to the collagen (CF) and elastic (EF) fibers. High magnification of the intracellular fibrils shows that they measure 30–80 A (inset). 4b: An individual fibrillar bundle with dense bodies at each extremity. *N*, nuclei; *EN*, endothelial cell; *A*, alveolus. (Figs. 2 and 4 from Kapanci, Y., Assimacopoulos, A., Irle, C., Zwahlen, A., & Gabbiani, G. (1974). "Contractile interstitial cells" in pulmonary alveolar septa: A possible regulator of ventilation/perfusion ratio?. Journal of Cell Biology, 60(2), 375–392. https://doi.org/10.1083/jcb.60.2.375. Copyright © 1974 Rockefeller University Press. All rights reserved.)

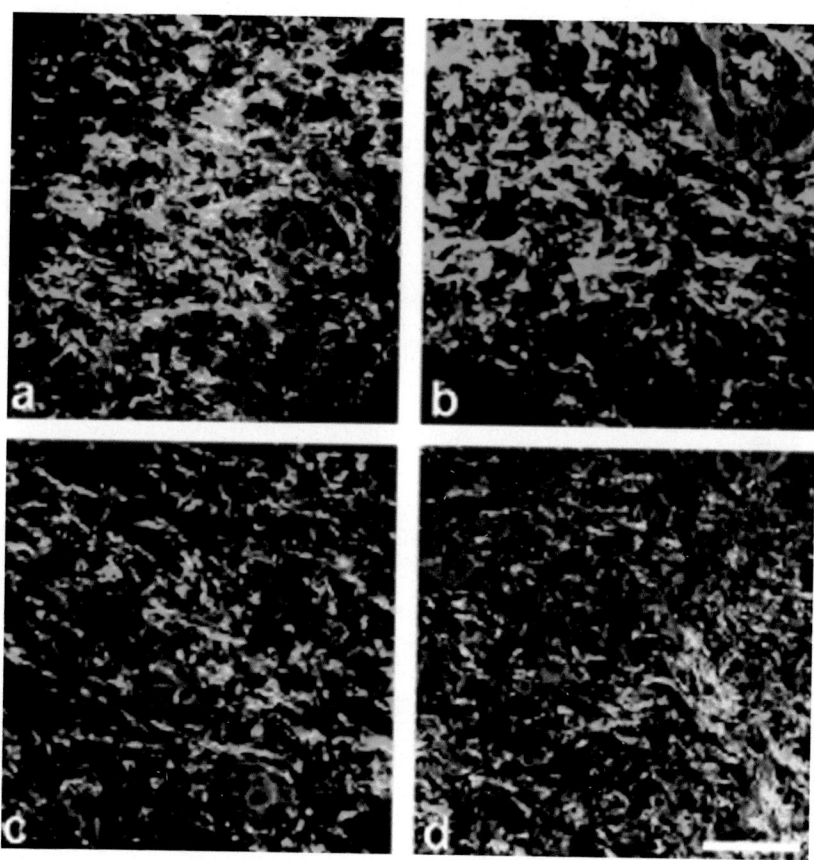

Fig. 4.5 α-SMA and ED-A fibronectin (FN) expression in granulation tissues at different time after wounding examined by confocal laser scanning microscopy. Sections of 4- (A and B), 7- (C), and 12-day-old (D) granulation tissue were double stained for total actin (A, *red*) or α-SMA (B–D, *red*) and ED-A FN (A–D, *green*). (A) Fibroblastic cells showing an important cytoplasmic staining for total actin are already present and interact with ED-A FN *(yellow* staining, corresponding to the overlay of *red* and *green* staining) which appear de novo in huge amounts as early as 4 d after wounding within the granulation tissue stroma. (B) At 4 d α-SMA is localized only in connection of SM-cells of small blood vessels, but not in connection with fibroblasts. (C) A 7-d-old wound tissue shows focal α-SMA staining of fibroblasts within an ED-A FN rich extracellular network. (D) 12-d-old granulation tissue fibroblastic cells show wide positivity for α-SMA. α-SMA and ED-A FN are colocalized (C and D; *yellow*); areas of colocalization are more abundant in 12- (D) than in 7-d-old (C) granulation tissue. Bar 50 μM. *(Fig. 1 from Serini, G., Bochaton-Piallat, M.-L., Ropraz, P., Geinoz, A., Borsi, L., Zardi, L., & Gabbiani, G. (1998). The fibronectin domain ED-A is crucial for myofibroblastic phenotype induction by transforming growth factor-β1. Journal of Cell Biology, 142(3), 873–881. https://doi.org/10.1083/jcb.142.3.873. Copyright © 1998 Rockefeller University Press. All rights reserved.)*

Celetta, Tomasek, Gabbiani, & Chaponnier, 2001). It has also been shown that γ-interferon inhibits myofibroblast differentiation and contractile activity indicating that the development of inflammation can influence the evolution of granulation tissue (Gagliano et al., 2020).

References

Bochaton-Piallat, M. L., Gabbiani, G., & Hinz, B. (2016). The myofibroblast in wound healing and fibrosis: Answered and unanswered questions. *F1000Res, 5.* https://doi.org/10.12688/f1000research.8190.1.

Darby, I., Skalli, O., & Gabbiani, G. (1990). Alpha-smooth muscle actin is transiently expressed by myofibroblasts during experimental wound healing. *Laboratory Investigation, 63*(1), 21–29.

Desmoulière, A., Redard, M., Darby, I., & Gabbiani, G. (1995). Apoptosis mediates the decrease in cellularity during the transition between granulation tissue and scar. *American Journal of Pathology, 146*(1), 56–66.

Fu, X., Kahlil, H., Boyer, J. G., Vagnozzi, R. J., Maliken, B. D., Sargent, M. A., ... Molkentin, J. D. (2018). Specialized fibroblast differentiated states underlie scar formation in the infarted mouse heart. *Journal of Clinical Investigation, 128*(5), 2127–2143. https://doi.org/10.1172/JCI98215.

Gabbiani, G., Chaponnier, C., & Hüttner, I. (1978). Cytoplasmic filaments and gap junctions in epithelial cells and myofibroblasts during wound healing. *Journal of Cell Biology, 76*(3), 561–568. https://doi.org/10.1083/jcb.76.3.561.

Gabbiani, G., Hirschel, B. J., Ryan, G. B., Statkov, P. R., & Majno, G. (1972). Granulation tissue as a contractile organ. A study of structure and function. *The Journal of Experimental Medicine, 135*(4), 719–734. https://doi.org/10.1084/jem.135.4.719.

Gabbiani, G., & Majno, G. (1972). Dupuytren's contracture: Fibroblast contraction? An ultrastructural study. *The American Journal of Pathology, 66*(1), 131–146.

Gabbiani, G., Ryan, G. B., & Majno, G. (1971). Presence of modified fibroblasts in granulation tissue and their possible role in wound contraction. *Experientia, 27*(5), 549–550. https://doi.org/10.1007/bf02147594.

Gagliano, T., Shah, K., Gargani, S., Lao, L., Alsaleem, M., Chen, J., ... Giamas, G. (2020). PIK3Cδ expression by fibroblasts promotes triple-negative breast cancer progression. *The Journal of Clinical Investigation, 130*(6), 3188–3204. https://doi.org/10.1172/jci128313.

Herrera, J., Henke, C. A., & Bitterman, P. B. (2018). Extracellular matrix as a driver of progressive fibrosis. *The Journal of Clinical Investigation, 128*(1), 45–53. https://doi.org/10.1172/jci93557.

Hinz, B., Celetta, G., Tomasek, J. J., Gabbiani, G., & Chaponnier, C. (2001). Alpha-smooth muscle actin expression upregulates fibroblast contractile activity. *Molecular Biology of the Cell, 12*(9), 2730–2741. https://doi.org/10.1091/mbc.12.9.2730.

Hinz, B., Phan, S. H., Thannickal, V. J., Prunotto, M., Desmoulière, A., Varga, J., ... Gabbiani, G. (2012). Recent developments in myofibroblast biology: Paradigms for connective tissue remodeling. *The American Journal of Pathology, 180*(4), 1340–1355.

Kang, S. H., Oh, S. Y., Lee, H. J., Kwon, T. G., Kim, J. W., Lee, S. T., ... Hong, S. H. (2021). Cancer-associated fibroblast subgroups showing differential promoting effect on HNSCC progression. *Cancers (Basel), 13*(4). https://doi.org/10.3390/cancers13040654.

Kapanci, Y., Assimacopoulos, A., Irle, C., Zwahlen, A., & Gabbiani, G. (1974). "Contractile interstitial cells" in pulmonary alveolar septa: A possible regulator of ventilation/perfusion ratio? *Journal of Cell Biology, 60*(2), 375–392. https://doi.org/10.1083/jcb.60.2.375.

Liao, Z., Tan, Z. W., Zhu, P., & Tan, N. S. (2019). Cancer-associated fibroblasts in tumor microenvironment—Accomplices in tumor malignancy. *Cellular Immunology, 343*, 103729. https://doi.org/10.1016/j.cellimm.2017.12.003.

Midgley, A. C., Woods, E. L., Jenkins, R. H., Brown, C., Khalid, U., Chavez, R., … Meran, S. (2020). Hyaluronidase-2 regulates RhoA signaling, myofibroblast contractility, and other key profibrotic myofibroblast functions. *The American Journal of Pathology, 190*(6), 1236–1255. https://doi.org/10.1016/j.ajpath.2020.02.012.

Patten, J., & Wang, K. (2021). Fibronectin in development and wound healing. *Advanced Drug Delivery Reviews, 170*, 353–368. https://doi.org/10.1016/j.addr.2020.09.005.

Pollard, T. D. (1976). Cytoskeletal functions of cytoplasmic contractile proteins. *Journal of Supramolecular Structure, 5*(3), 317–334. https://doi.org/10.1002/jss.400050306.

Qadri, M., Jay, G. D., Zhang, L. X., Richendrfer, H., Schmidt, T. A., & Elsaid, K. A. (2020). Proteoglycan-4 regulates fibroblast to myofibroblast transition and expression of fibrotic genes in the synovium. *Arthritis Research & Therapy, 22*(1), 113. https://doi.org/10.1186/s13075-020-02207-x.

Santucci, M., Borgognoni, L., Reali, U. M., & Gabbiani, G. (2001). Keloids and hypertrophic scars of Caucasians show distinctive morphologic and immunophenotypic profiles. *Virchows Archiv, 438*(5), 457–463. https://doi.org/10.1007/s004280000335.

Sappino, A. P., Masouyé, I., Saurat, J. H., & Gabbiani, G. (1990). Smooth muscle differentiation in scleroderma fibroblastic cells. *The American Journal of Pathology, 137*(3), 585–591.

Schleip, R., Gabbiani, G., Wilke, J., Naylor, I., Hinz, B., Zorn, A., … Klingler, W. (2019). Fascia is able to actively contract and may thereby influence musculoskeletal dynamics: A histochemical and mechanographic investigation. *Frontiers in Physiology, 10*, 336. https://doi.org/10.3389/fphys.2019.00336.

Serini, G., Bochaton-Piallat, M.-L., Ropraz, P., Geinoz, A., Borsi, L., Zardi, L., & Gabbiani, G. (1998). The fibronectin domain ED-A is crucial for myofibroblastic phenotype induction by transforming growth factor-β1. *Journal of Cell Biology, 142*(3), 873–881. https://doi.org/10.1083/jcb.142.3.873.

Skalli, O., Ropraz, P., Trzeciak, A., Benzonana, G., Gillessen, D., & Gabbiani, G. (1986). A monoclonal antibody against alpha-smooth muscle actin: A new probe for smooth muscle differentiation. *The Journal of Cell Biology, 103*(6 Pt 2), 2787–2796. https://doi.org/10.1083/jcb.103.6.2787.

Tomasek, J. J., Gabbiani, G., Hinz, B., Chaponnier, C., & Brown, R. A. (2002). Myofibroblasts and mechano-regulation of connective tissue remodelling. *Nature Reviews. Molecular Cell Biology, 3*(5), 349–363. https://doi.org/10.1038/nrm809.

Zampieri, F., Coen, M., & Gabbiani, G. (2014). The prehistory of the cytoskeleton concept. *Cytoskeleton (Hoboken), 71*(8), 464–471. https://doi.org/10.1002/cm.21177.

CHAPTER 5

Conclusions and perspectives

We are not far from the 50th anniversary of the publication of the first paper describing the myofibroblast (Gabbiani, Ryan, & Majno, 1971) and an enormous progress has been made on the understanding of the biological features of this complex cell as well as of his involvement in many important physiological and pathological events, such as tension regulation in the normal pulmonary alveolus and in fibrosis of heart, lungs, liver, and kidney (Bochaton-Piallat, Gabbiani, & Hinz, 2016; Tomasek, Gabbiani, Hinz, Chaponnier, & Brown, 2002). Moreover, the description of the myofibroblast contributed to the morphological and functional definition of microfilaments and thus to the characterization of the family of distinct filamentous structures (microfilaments, intermediate filaments, and microtubules) composing the cytoskeleton (for a review see Zampieri, Coen, & Gabbiani, 2014). However, many aspects of myofibroblast biology and of his pathological role remain unknown. Among the mechanisms involved in myofibroblast activities, the mechanical aspect has been intensively investigated and has allowed decisive advances in the understanding of the crosstalk between force generation and biological activities in connective tissue and in the interaction between mesenchymal and epithelial cells (Bochaton-Piallat et al., 2016). Among growth factors, TGF-β1, either produced by inflammatory cells or by the myofibroblast itself has been shown to be essential in the evolution of myofibroblast-dependent phenomena. Unfortunately, due to the complex regulation of TGF-β1 expression, it has not yet been possible to obtain relevant results using direct inhibitors, but clinical trials that target TGF-β1 are under way with the hope of obtaining new encouraging data (Bochaton-Piallat et al., 2016; Weng, Mertens, Gressner, & Dooley, 2007). EDA fibronectin, produced at least in part by the myofibroblast, appears necessary for myofibroblast differentiation, suggesting that it could represent a therapeutic target (Serini et al., 1998). Fibroblast and myofibroblast phenotypic heterogeneity start to be explored and should furnish, in addition to a new classification of distinct fibroblastic phenotypes, new tools for therapeutic strategies

Wound Healing, Fibrosis, and the Myofibroblast
https://doi.org/10.1016/B978-0-323-90546-6.00009-5

(Bochaton-Piallat et al., 2016). In particular, it would be important to distinguish distinct fibroblastic phenotypes involved in myofibroblast differentiation, collagen and/or α-SMA synthesis; this would facilitate an early approach to therapeutic studies during myofibroblast-dependent pathological processes. Reduction of α-SMA expression by myofibroblast has been shown experimentally to reduce wound contraction and should be explored as another therapeutic target (Hinz, Gabbiani, & Chaponnier, 2002). Another aspect of myofibroblast biology interesting several laboratories is the production by these cells of soluble substances, which can influence the behavior of inflammatory, connective tissue and epithelial cells, thus establishing a cross talk between tissue components that is at the basis of processes such as fibrosis evolution, cancer cell invasion or metastasis formation (Bochaton-Piallat et al., 2016). Recently, it has been shown that fibroblast-expressed kinases, in particular PIK3Cδ, can work as tumor suppressor in a murine model of triple negative breast cancer suggesting PIK3Gδ as a potential therapeutic target (Gagliano et al., 2020).

As discussed above, myofibroblast function is highly dependent on its interaction with extracellular matrix. A recent report indicates that collagen type V, despite a low degree of expression, exerts a regulatory function on experimental heart scar evolution, indicating a new avenue of therapeutic strategy in heart fibrosis control (Yokota et al., 2020).

Much work remains to do in order to understand many aspects of myofibroblast biology and role in pathological phenomena, but, in view of the undergoing experimental efforts, we are confident that the next few years will bring new interesting discoveries in these fields as well as new therapeutic tools for myofibroblast-dependent diseases.

References

Bochaton-Piallat, M. L., Gabbiani, G., & Hinz, B. (2016). The myofibroblast in wound healing and fibrosis: Answered and unanswered questions. *F1000Res, 5.* https://doi.org/10.12688/f1000research.8190.1.

Gabbiani, G., Ryan, G. B., & Majno, G. (1971). Presence of modified fibroblasts in granulation tissue and their possible role in wound contraction. *Experientia, 27*(5), 549–550. https://doi.org/10.1007/bf02147594.

Gagliano, T., Shah, K., Gargani, S., Lao, L., Alsaleem, M., Chen, J., … Giamas, G. (2020). PIK3Cδ expression by fibroblasts promotes triple-negative breast cancer progression. *The Journal of Clinical Investigation, 130*(6), 3188–3204. https://doi.org/10.1172/jci128313.

Hinz, B., Gabbiani, G., & Chaponnier, C. (2002). The NH2-terminal peptide of alpha-smooth muscle actin inhibits force generation by the myofibroblast in vitro and in vivo. *The Journal of Cell Biology, 157*(4), 657–663. https://doi.org/10.1083/jcb.200201049.

Serini, G., Bochaton-Piallat, M.-L., Ropraz, P., Geinoz, A., Borsi, L., Zardi, L., & Gabbiani, G. (1998). The fibronectin domain ED-A is crucial for myofibroblastic phenotype induction by transforming growth factor-β1. *Journal of Cell Biology, 142*(3), 873–881. https://doi.org/10.1083/jcb.142.3.873.

Tomasek, J. J., Gabbiani, G., Hinz, B., Chaponnier, C., & Brown, R. A. (2002). Myofibroblasts and mechano-regulation of connective tissue remodelling. *Nature Reviews. Molecular Cell Biology, 3*(5), 349–363. https://doi.org/10.1038/nrm809.

Weng, H., Mertens, P. R., Gressner, A. M., & Dooley, S. (2007). IFN-gamma abrogates profibrogenic TGF-beta signaling in liver by targeting expression of inhibitory and receptor Smads. *Journal of Hepatology, 46*(2), 295–303. https://doi.org/10.1016/j.jhep.2006.09.014.

Yokota, T., McCourt, J., Ma, F., Ren, S., Li, S., Kim, T.-H., ... Deb, A. (2020). Type V collagen in scar tissue regulates the size of scar after heart injury. *Cell, 182*(3), 545–562.e523. https://doi.org/10.1016/j.cell.2020.06.030.

Zampieri, F., Coen, M., & Gabbiani, G. (2014). The prehistory of the cytoskeleton concept. *Cytoskeleton (Hoboken), 71*(8), 464–471. https://doi.org/10.1002/cm.21177.

Index

Note: Page numbers followed by *f* indicate figures.